T0324179

Atlantis Studies in Scientific Computing in Electromagnetics

Volume 2

Series editor

Wil Schilders, Dept Mathematics & Comp Sci, Eindhoven Univ of Tech, Eindhoven, The Netherlands

This series contains volumes on scientific computing for a wide range of electromagnetics problems. The electronics industry, in a very broad sense, is at the forefront of innovation, and our daily life is very much dependent on achievements in this area. These are mainly enabled by rapid developments in sophisticated virtual design environments, numerical methods being at the core of these. Volumes in the series provide details on the modeling, analysis and simulation of problems, as well as on the design process of robust solution methods. Applications range from simple benchmark problems to industrial problems at the forefront of current research and development.

For more information on this series and our other book series, please visit our website: www.atlantis-press.com

Atlantis Press
8 square des Bouleaux
75019 Paris, France

More information about this series at http://www.springer.com/series/13301

N. Banagaaya · G. Alì · W.H.A. Schilders

Index-aware Model Order Reduction Methods

Applications to Differential-Algebraic Equations

ATLANTIS
PRESS

N. Banagaaya
Computational Methods in Systems and Control Theory
MPI for Dynamics of Complex Technical Systems
Magdeburg
Germany

W.H.A. Schilders
Department of Mathematics and Computer Science
Eindhoven University of Technology
Eindhoven
The Netherlands

G. Alì
Department of Physics
University of Calabria
Arcavacata di Rende
Cosenza
Italy

ISSN 2352-0590 ISSN 2352-0604 (electronic)
Atlantis Studies in Scientific Computing in Electromagnetics
ISBN 978-94-6239-188-8 ISBN 978-94-6239-189-5 (eBook)
DOI 10.2991/978-94-6239-189-5

Library of Congress Control Number: 2016932327

© Atlantis Press and the author(s) 2016
This book, or any parts thereof, may not be reproduced for commercial purposes in any form or by any means, electronic or mechanical, including photocopying, recording or any information storage and retrieval system known or to be invented, without prior permission from the Publisher.

Printed on acid-free paper

Contents

List of Figures

List of Tables

Chapter 1
Introduction

The main aim of this book is to discuss model order reduction (MOR) methods for linear coefficients differential algebaric equations (DAEs). Our main discussion will focus on the index-aware model order reduction method (IMOR) and its invariants proposed in [1, 2, 4, 5]. Modeling of sophisticated applications such as problems arising from nanoelectronics, electrical networks, multibody systems, aerospace engineering, chemical processes, computational fluid dynamics (CFD), gas transport networks, see [10–12, 15, 23, 35], can lead to DAEs that are either linear or nonlinear in state space. These systems are always large in dimension compared to the number of inputs and outputs due to the need of highly accurate models which must be closer to the physical model. This is always done by using a very fine mesh grid which can lead to systems of a very large dimension.

1.1 Real-Life Applications of DAEs

DAEs appear in many fields as mentioned earlier. In this section, we present some of the applications of DAEs in the real-world. The mathematical models of these applications can be written as

$$\mathbf{E}\dot{\boldsymbol{x}}(t) = \mathbf{A}\boldsymbol{x}(t) + \mathbf{B}\boldsymbol{u}(t), \quad \boldsymbol{x}(0) = \boldsymbol{x}_0, \tag{1.1.1a}$$

$$\boldsymbol{y}(t) = \mathbf{C}^T\boldsymbol{x}(t), \tag{1.1.1b}$$

where $\mathbf{E} \in \mathbb{R}^{n\times n}$ is singular, $\mathbf{A} \in \mathbb{R}^{n\times n}$, $\mathbf{B} \in \mathbb{R}^{n\times m}$, $\mathbf{C} \in \mathbb{R}^{n\times \ell}$, and $\boldsymbol{u}(t) \in \mathbb{R}^m$ is the input vector, $\boldsymbol{y}(t) \in \mathbb{R}^\ell$ is the output vector, $\boldsymbol{x}_0 \in \mathbb{R}^n$ is the initial value.

© Atlantis Press and the author(s) 2016

N. Banagaaya et al., *Index-aware Model Order Reduction Methods*,
Atlantis Studies in Scientific Computing in Electromagnetics 2,
DOI 10.2991/978-94-6239-189-5_1

1.1.1 Electrical Network Problems

Many electrical circuit systems can be described by DAEs of the form (1.1.1). This is due to the fact that, the most commonly used method in electrical circuit networks design is the modified nodal analysis (MNA). This approach leads to a DAE when modeling a network involving resistor networks such as RLC networks, i.e., Resistor-Inductor-Capacitor network, RC networks, i.e., Resistor-Capacitor network, RL networks, i.e., Resistor-Inductor network and so on. For illustration, we consider only RLC and RC networks.

(i) **RLC network**. Consider a linear RLC electric network, that is, a network which connects linear capacitors, inductors and resistors, and current sources, $\boldsymbol{v}(t) \in \mathbb{R}^{n_V}$ and $\boldsymbol{\imath}(t) \in \mathbb{R}^{n_I}$. The unknowns which describe the network are the node potentials $\boldsymbol{e}(t) \in \mathbb{R}^n$, and the currents through inductors $\boldsymbol{\jmath}_L(t) \in \mathbb{R}^{n_L}$. Following the formalism of modified nodal analysis (MNA) [18, 26], we introduce: the incidence matrices $\boldsymbol{A}_C \in \mathbb{R}^{n \times n_C}$, $\boldsymbol{A}_L \in \mathbb{R}^{n \times n_L}$ and $\boldsymbol{A}_R \in \mathbb{R}^{n \times n_G}$, which describe the branch-node relationships for capacitors, inductors and resistors; the incidence matrices $\boldsymbol{A}_V \in \mathbb{R}^{n \times n_V}$ and $\boldsymbol{A}_I \in \mathbb{R}^{n \times n_I}$, which describe this relationship for voltage and current sources, respectively. Then, the DAE model for the RLC network with unknown $\boldsymbol{x} = (\boldsymbol{e}, \boldsymbol{\jmath}_L, \boldsymbol{\jmath}_V)^{\mathrm{T}}$ is given by

$$\begin{pmatrix} \boldsymbol{A}_C \boldsymbol{C} \boldsymbol{A}_C^{\mathrm{T}} & \mathbf{0} & \mathbf{0} \\ \mathbf{0} & \boldsymbol{L} & \mathbf{0} \\ \mathbf{0} & \mathbf{0} & \mathbf{0} \end{pmatrix} \frac{d\boldsymbol{x}}{dt} = \begin{pmatrix} -\boldsymbol{A}_R \boldsymbol{G} \boldsymbol{A}_R^{\mathrm{T}} & -\boldsymbol{A}_L & -\boldsymbol{A}_V \\ \boldsymbol{A}_L^{\mathrm{T}} & \mathbf{0} & \mathbf{0} \\ \boldsymbol{A}_V^{\mathrm{T}} & \mathbf{0} & \mathbf{0} \end{pmatrix} \boldsymbol{x} + \begin{pmatrix} -\boldsymbol{A}_I & \mathbf{0} \\ \mathbf{0} & \mathbf{0} \\ \mathbf{0} & -\mathrm{I} \end{pmatrix} \begin{pmatrix} \boldsymbol{\imath} \\ \boldsymbol{v} \end{pmatrix}. \tag{1.1.2}$$

(ii) **RC network**. The RC model can be derived from that of the RLC model (1.1.2) by simply eliminating the inductor currents $\boldsymbol{\jmath}_L$. Then, the DAE model of the RC network with unknown $\boldsymbol{x} = (\boldsymbol{e}, \boldsymbol{\jmath}_V)^{\mathrm{T}}$ is given by

$$\begin{pmatrix} \boldsymbol{A}_C \boldsymbol{C} \boldsymbol{A}_C^{\mathrm{T}} & \mathbf{0} \\ \mathbf{0} & \mathbf{0} \end{pmatrix} \frac{d\boldsymbol{x}}{dt} = \begin{pmatrix} -\boldsymbol{A}_R \boldsymbol{G} \boldsymbol{A}_R^{\mathrm{T}} & -\boldsymbol{A}_V \\ \boldsymbol{A}_V^{\mathrm{T}} & \mathbf{0} \end{pmatrix} \boldsymbol{x} + \begin{pmatrix} -\boldsymbol{A}_I & \mathbf{0} \\ \mathbf{0} & -\mathrm{I} \end{pmatrix} \begin{pmatrix} \boldsymbol{\imath} \\ \boldsymbol{v} \end{pmatrix}. \tag{1.1.3}$$

We can observe that (1.1.2) and (1.1.3) are DAEs of the form (1.1.1a).

1.1.2 Computational Fluid Dynamics Problems

(i) **Supersonic Inlet flow example**. This example originates from [23]. Consider an unsteady flow through a supersonic diffuser as shown in Fig. 1.1. The diffuser operates at a nominal Mach number of 2.2, however it is subject to perturbations in the incoming flow, which may be due to atmospheric variations. In nominal operation, there is a strong shock downstream of the diffuser throat, as can be seen from the Mach contours plotted in Fig. 1.1. Incoming disturbances can

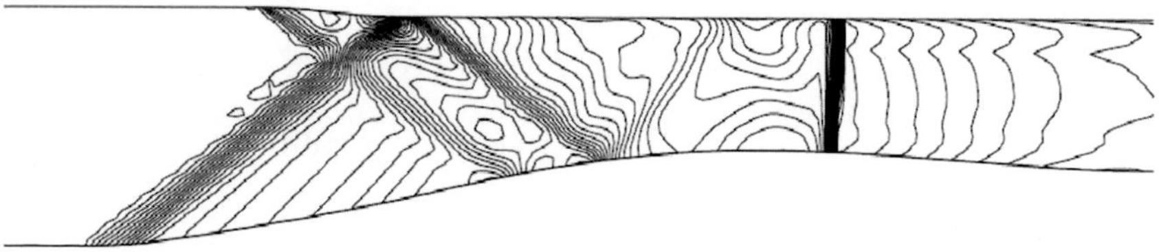

Fig. 1.1 Steady-state mach contours inside diffuser

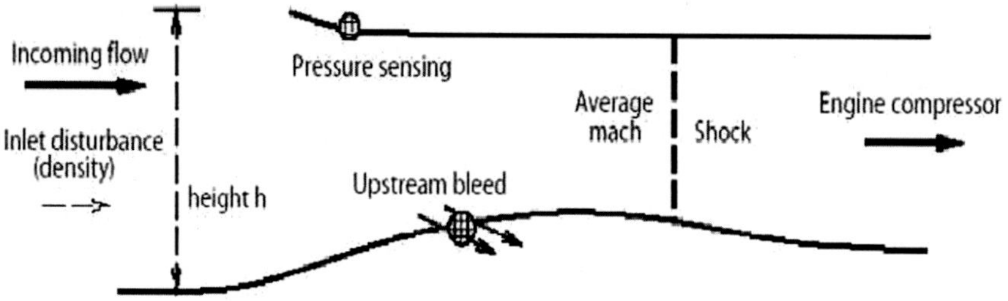

Fig. 1.2 Supersonic diffuser active flow control problem setup

cause the shock to move forward towards the throat. When the shock sits at the throat, the inlet is unstable, since any disturbance that moves the shock slightly upstream will cause it to move forward rapidly, leading to unstart of the inlet. This is extremely undesirable, since unstart results in a large loss of thrust. In order to prevent unstart from occurring, one option is to actively control the position of the shock. This control may be effected through flow bleeding upstream of the diffuser throat. In order to derive effective active control strategies, it is imperative to have low-order models which accurately capture the relevant dynamics. Figure 1.2 presents the schematic of the actuation mechanism. Incoming flow with possible disturbances enters the inlet and is sensed using pressure sensors. The controller then adjusts the bleed upstream of the throat in order to control the position of the shock and to prevent it from moving upstream. In simulations, it is difficult to automatically determine the shock location. The average Mach number at the diffuser throat provides an appropriate surrogate that can be easily computed. There are several transfer functions of interest in this problem. The shock position will be controlled by monitoring the average Mach number at the diffuser throat. The reduced-order model must capture the dynamics of this output in response to two inputs: the incoming flow disturbance and the bleed actuation. In addition, total pressure measurements at the diffuser wall are used for sensing. The response of this output to the two inputs must also be captured. This problem is modeled using an unsteady, two-dimensional flow of an inviscid, compressible fluid which is governed by the Euler equations. The usual statements of mass, momentum, and energy can be written in integral form as

$$\begin{aligned} \frac{\partial}{\partial t} \iint \rho \, \mathrm{d}V + \oint \rho \mathbf{Q}.\mathbf{dA} &= 0 \\ \frac{\partial}{\partial t} \iint \rho \mathbf{Q} \, \mathrm{d}V + \oint \rho \mathbf{Q}(\mathbf{Q}.\mathbf{dA}) + \oint p\mathbf{dA} &= 0 \\ \frac{\partial}{\partial t} \iint \rho E \, \mathrm{d}V + \oint \rho H(\mathbf{Q}.\mathbf{dA}) + \oint p\mathbf{Q}.\mathbf{dA} &= 0, \end{aligned} \tag{1.1.4}$$

where ρ, $\mathbf{Q}$, H, E, and p denote density, flow velocity, total enthalpy, energy, and pressure, respectively. The CFD formulation for this problem uses a finite volume method and is described fully in Lassaux [22]. The unknown flow quantities used are the density, streamwise velocity component, normal velocity component, and enthalpy at each point in the computational grid. Note that the local flow velocity components q and $q^{\perp}$ are defined using a streamline computational grid that is computed for the steady-state solution. q is the projection of the flow velocity on the meanline direction of the grid cell, and q is the normal-to-meanline component. To simplify the implementation of the integral energy equation, total enthalpy is also used in place of energy. The vector of unknowns at each node i is therefore $\boldsymbol{x}_i = \left[\rho_i, q_i, q_i^{\perp}, H_i\right]^{\mathrm{T}}$. Two physically different kinds of boundary conditions exist: inflow/outflow conditions, and conditions applied at a solid wall. At a solid wall, the usual no-slip condition of zero normal flow velocity is easily applied as $q^{\perp} = 0$. In addition, we will allow for mass addition or removal (bleed) at various positions along the wall. The bleed condition is also easily specified. We set $q^{\perp} = \frac{\dot{m}}{\rho}$, where $\dot{m}$ is the specified mass flux per unit length along the bleed slot. At inflow boundaries, Riemann boundary conditions are used. For the diffuser problem considered here, all inflow boundaries are supersonic, and hence we impose inlet vorticity, entropy and Riemanns invariants. At the exit of the duct, we impose outlet pressure.
The two-dimensional integral Euler equations are linearized about the steady state solution to obtain a semi-explicit DAE of index-1 of the form (1.1.1) with system matrices

$$\mathbf{E} = \begin{pmatrix} \mathbf{E}_{11} & \mathbf{E}_{12} \\ \mathbf{0} & \mathbf{0} \end{pmatrix}, \ \mathbf{A} = \begin{pmatrix} \mathbf{A}_{11} & \mathbf{A}_{12} \\ \mathbf{A}_{21} & \mathbf{A}_{22} \end{pmatrix}, \ \mathbf{B} = \begin{pmatrix} \mathbf{B}_1 \\ \mathbf{B}_2 \end{pmatrix}, \ \mathbf{C} = \begin{pmatrix} \mathbf{C}_1 \\ \mathbf{C}_2 \end{pmatrix}, \ \boldsymbol{x} = \begin{pmatrix} \boldsymbol{x}_1 \\ \boldsymbol{x}_2 \end{pmatrix}, \tag{1.1.5}$$

where $\mathbf{E}_{11} \in \mathbb{R}^{n_1 \times n_1}$ and $\mathbf{A}_{21}\mathbf{E}_{11}^{-1}\mathbf{E}_{12} - \mathbf{A}_{22} \in \mathbb{R}^{n_2 \times n_2}$ are non-singular matrices due to index-1 property and $n = n_1 + n_2$ is the dimension of the DAE.

(ii) **Semidiscretized Stokes equation**. In this section, we present the semidiscretized Stokes equation originating from [35]. Consider the instationary Stokes equation describing the flow of an incompressible fluid

$$\frac{\partial \boldsymbol{v}}{\partial t} = \Delta \boldsymbol{v} - \nabla \boldsymbol{p} + \boldsymbol{f}, \quad (\zeta, t) \in \Omega \times (0, \mathrm{T}) \qquad (1.1.6)$$
$$0 = \mathrm{div} v,$$

with appropriate initial condition and boundary condition. Here $\boldsymbol{v}(\zeta, t) \in \mathbb{R}^d$ is the velocity vector (d = 2 or 3 is the dimension of the spatial domain), $\boldsymbol{p}(\zeta, t) \in \mathbb{R}$ is the pressure, $\boldsymbol{f}(\zeta, t) \in \mathbb{R}$ is the vector of external forces, $\Omega \in \mathbb{R}^d$ is a bounded open domain and $\mathrm{T} > 0$ is the endpoint of the time interval. The spatial discretization of the Stokes equation (1.1.6) by either the finite difference or finite element methods on a uniform staggered grid leads to a DAE of the form (2.3.1) with system matrices:

$$\mathbf{E} = \begin{pmatrix} \mathbf{E}_{11} & \mathbf{0} \\ \mathbf{0} & \mathbf{0} \end{pmatrix}, \ \mathbf{A} = \begin{pmatrix} \mathbf{A}_{11} & \mathbf{A}_{12} \\ \mathbf{A}_{12}^{\mathrm{T}} & \mathbf{0} \end{pmatrix}, \ \mathbf{B} = \begin{pmatrix} \mathbf{B}_1 \\ \mathbf{B}_2 \end{pmatrix}, \ \mathbf{C} = \begin{pmatrix} \mathbf{C}_1 \\ \mathbf{C}_2 \end{pmatrix}, \ \boldsymbol{x} = \begin{pmatrix} v_h \\ \boldsymbol{p}_h \end{pmatrix}, \qquad (1.1.7)$$

where $v_h \in \mathbb{R}^{n_1}$ and $\boldsymbol{p}_h \in \mathbb{R}^{n_2}$ are the semidiscretized vectors of velocity and pressure, respectively, see [35]. The matrix $\mathbf{E}_{11} \in \mathbb{R}^{n_1 \times n_1}$ is a nonsingular matrix, but for this case $\mathbf{E}_{11} = \mathbf{I}$, $\mathbf{A}_{11} \in \mathbb{R}^{n_1 \times n_1}$ is the discrete Laplace operator, $-\mathbf{A}_{12} \in \mathbb{R}^{n_1 \times n_2}$ and $-\mathbf{A}_{12}^{\mathrm{T}} \in \mathbb{R}^{n_2 \times n_1}$ are, the discrete gradient and divergence operators, respectively. Due to the non-uniqueness of the pressure, the matrix $\mathbf{A}_{12}$ has a rank defect one. In this case, instead of $\mathbf{A}_{12}$ we can take a full column rank matrix obtained from $\mathbf{A}_{12}$ by discarding the last column. Therefore, in the following we will assume without loss of generality that $\mathbf{A}_{12}$ has full column rank. In this case system with matrix coefficients (1.1.7) is of index-2. The matrices $\mathbf{B}_1 \in \mathbb{R}^{n_1 \times m}$, $\mathbf{B}_2 \in \mathbb{R}^{n_2 \times m}$ and the control input $\boldsymbol{u}(t) \in \mathbb{R}^m$ are the resulting from the boundary condition and external forces, the output $\boldsymbol{y}(t) \in \mathbb{R}^\ell$ is the vector of interest. The order $n = n_1 + n_2$ of system (1.1.7) depends on the level of refinement of the discretization and is usually very large, whereas the number m of inputs and the number ℓ of outputs are typically small.

1.1.3 Constrained Mechanical Problems

This example originates from [35]. We consider the holonomically constrained damped mass-spring system as illustrated in Fig. 1.3. The ith mass of weight m_i is connected to the $(i+1)$st mass by a spring and a damper with constants k_i and d_i, respectively, and also to the ground by a spring and a damper with constants k_i and δ_i, respectively. Additionally, the first mass is connected to the last one by a rigid bar and it is influenced by the control $\boldsymbol{u}(t)$. The vibration of this system is described by a DAE of the form (1.1.1) with system matrices

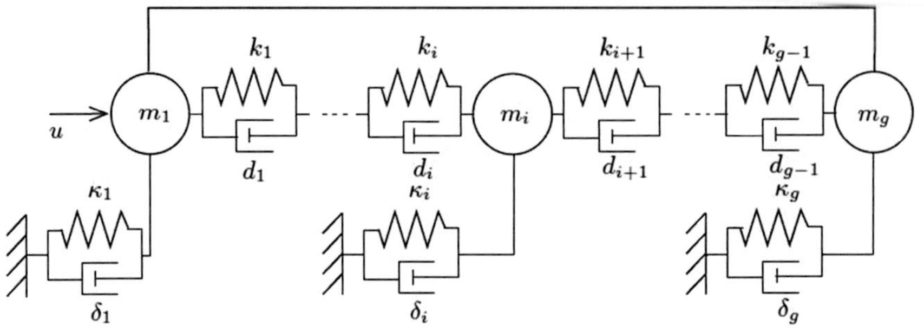

Fig. 1.3 A damped mass-spring system with a holonomic constraint

$$\mathbf{E} = \begin{pmatrix} \mathrm{I} & \mathbf{0} & \mathbf{0} \\ \mathbf{0} & \mathbf{M} & \mathbf{0} \\ \mathbf{0} & \mathbf{0} & \mathbf{0} \end{pmatrix}, \mathbf{A} = \begin{pmatrix} 0 & \mathrm{I} & \mathbf{0} \\ \mathbf{K} & \mathbf{D} & -\mathbf{G}^{\mathrm{T}} \\ \mathbf{G} & \mathbf{0} & \mathbf{0} \end{pmatrix}, \mathbf{B} = \begin{pmatrix} \mathbf{0} \\ \mathbf{B}_2 \\ \mathbf{0} \end{pmatrix}, \mathbf{C} = \begin{pmatrix} \mathbf{C}_1 \\ \mathbf{0} \\ \mathbf{0} \end{pmatrix}, \boldsymbol{x}(t) = \begin{pmatrix} \boldsymbol{p}(t) \\ \boldsymbol{v}(t) \\ \boldsymbol{\lambda}(t) \end{pmatrix}, \tag{1.1.8}$$

where $\boldsymbol{p}(t) \in \mathbb{R}^g$ is the position vector, $\boldsymbol{v}(t) \in \mathbb{R}^g$ is the velocity vector, $\boldsymbol{\lambda}(t) \in \mathbb{R}$ is the Lagrange multiplier, $\mathbf{M} = \mathrm{diag}(m_1, \ldots, m_g)$ is the mass matrix, $\mathbf{D}$ and $\mathbf{K}$ are the tridiagonal damping and stiffness matrices, $\mathbf{G} = [1, 0, \ldots, 0, -1] \in \mathbb{R}^{1\times g}$ is the constraint matrix, $\mathbf{B}_2 = \boldsymbol{e}_1$ and $\mathbf{C}_1 = [\boldsymbol{e}_1, \boldsymbol{e}_2, \ldots, \boldsymbol{e}_{g-1}]^{\mathrm{T}}$. Here $\boldsymbol{e}_i$ denotes the ith column of the identity matrix I_g. Thus the system is of dimension $n = 2g + 1$. According to [35], system (1.1.8) is of index-3 since $\mathbf{G}$ is a full row rank.

1.2 Why MOR for DAEs?

A typical feature of DAEs is that they lead to state space descriptions of high dimension in which the coefficient of the first order derivative is a singular matrix. If the initial condition is inconsistent or when the smoothness of the input does not correspond to the index of the DAE, currently available MOR techniques may lead to inaccurate reduced-order models. These reduced-order models may lead to wrong solutions that do not adequately represent the hidden truly fast modes or are very difficult to solve numerically.

However, it happens that he conventional MOR methods [3, 9, 31] cannot be applied immediately especially to higher index DAEs because they were originally developed for system possessing zero initial condition. There has been attempts of using MOR methods, such as PRIMA method [28], for ODE systems in descriptor form to be used to reduced DAEs but these methods treat DAEs as ordinary differential equations (ODEs). Moreover, sometimes these methods lead to ODEs reduced-order models even if the original model is a DAE. This may lead to loss of their mathematical properties. As a consequence, new concepts were needed to provide reliable reduced-order models for DAEs. In the new approach, DAEs must first be decoupled into differential and algebraic parts before applying any MOR technique. This observation has led to the development of new methods specifically for DAEs, see [13, 14, 16, 19, 35] and to some extent the modification of the existing MOR methods, see [16, 35]. Most of these recently developed methods are application-based, and some are more general. In [35], the authors proposed the most successful MOR method for DAEs, known as the balanced truncation method for descriptor systems. However this method is computationally expensive since it involves solving four Lyapunov equations and it relies on the construction of spectral projectors which are well known to be numerically infeasible. Hence, the application of the balanced truncation method is limited to DAEs with special structures.

In this book, we discuss a computationally cheaper way of decoupling and reduction of DAEs. This decoupling procedure relies on the framework of a special projector and matrix chain for differential algebraic systems, enabling a decomposition of the DAEs into separate differential and algebraic parts as introduced by März in [25]. However, the März decoupling procedure leads to much larger decoupled system of dimension $n(\mu + 1)$, where n and μ is the dimension and the tractability index of the DAE, respectively, and does not preserve its stability. In [1, 2], the März decoupling procedure is modified by using the special bases of the corresponding projectors, thus preserving the dimension and stability of the equivalnet decoupled system. Having performed this separation, different reduction methods can be used to each of these parts. For the differential part, one can use conventional MOR methods, while for the algebraic part new MOR methods for algebraic systems have to be developed. This procedure led to a MOR method for DAEs which was called the Index-aware MOR method, abbreviated as IMOR method [1, 2]. This method is illustrated in Fig. 1.4. This method is very robust and leads to simple reduced-order models even for higher index DAEs. However, the IMOR method has an inherited limitation of matrix inversion which makes it computationally very expensive. In [5], the authors developed an implicit version of the IMOR method which they called

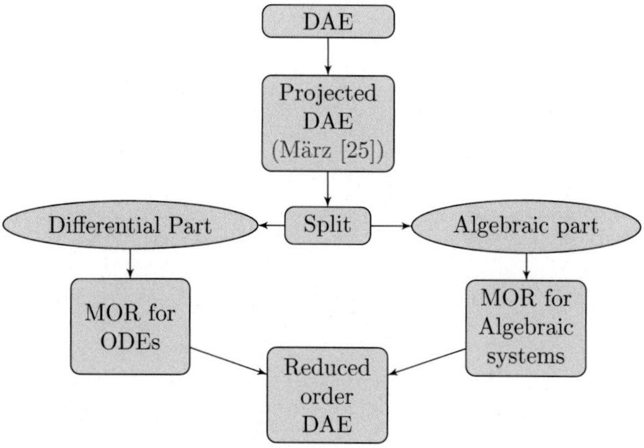

Fig. 1.4 IMOR methods procedure

the implicit IMOR (IIMOR) method. This method is computationally cheaper than its explicit counter part. However, experiments show that the IMOR method is more accurate thus, one needs to trade off between complexity and accuracy. Using the derived decoupled systems as a result of the IMOR and IIMOR methods, we can be able to analyse the limitations of the conventional MOR methods. We observed that sometimes conventional MOR methods can lead to accurate reduced-order models even for higher index DAEs, if and only if the consistent initial condition does not depend on the derivatives of the input data. The explicit and implicit decoupling procedures are also advantageous for solving DAEs more efficiently numerically since they enable one to use the conventional ODE integration methods to solve higher index DAEs.

Other well known tools to investigate DAEs are the transformation into Kronecker normal form and the decoupling by means of Drazin inverses and spectral projectors. These tools are very accurate but they are numerically infeasible and can not be generalized to variable coefficient linear DAEs and nonlinear DAEs. However, our decoupling procedures used in both IMOR and IIMOR methods, relies on the matrix and projector chain approach introduced by März [25] which also applies to the case of general variable coefficient equations, see [17]. Hence the IMOR and IIMOR methods can be extended to variable coefficient linear DAEs and nonlinear DAEs.

Chapter 2
Differential-Algebraic Equations

In this chapter, we introduce the differential algebraic equations which we abbreviate as DAEs. DAEs arise in a variety of applications such as modelling constrained multibody systems, electrical networks, aerospace engineering, chemical processes, computational fluid dynamics, gas transport networks, see [10–12, 35]. Therefore their analysis and numerical treatment plays an important role in modern mathematics. In many articles, DAEs are also called singular systems [11], descriptor systems [12, 35, 36], generalized state space systems [35], semi-state systems, degenerated systems, constrained systems, implicit systems but in most literatures they are called DAEs [24, 25, 29]. In this book, we shall also call them DAEs.

2.1 What are DAEs?

Consider an explicit ordinary differential equation (ODE),

$$\dot{\boldsymbol{x}} = \boldsymbol{f}(t, \boldsymbol{x}), \tag{2.1.1}$$

where $\dot{\boldsymbol{x}} = \frac{\mathrm{d}x}{\mathrm{d}t}$ and $\boldsymbol{x} \in \mathbb{R}^n, \boldsymbol{f} : \mathbb{R} \times \mathbb{R}^n \to \mathbb{R}^n$. In general the first order ODE system can be written in implicit form:

$$\boldsymbol{F}(t, \boldsymbol{x}, \dot{\boldsymbol{x}}) = \boldsymbol{0}. \tag{2.1.2}$$

According to [32], if the Jacobian matrix $\frac{\partial F}{\partial \dot{x}}$ is nonsingular then it is possible to solve (2.1.2) for $\dot{\boldsymbol{x}}$ in order to obtain an ODE (2.1.1). However, if $\frac{\partial F}{\partial \dot{x}}$ is singular, this is no longer possible and the solution $\boldsymbol{x}$ has to satisfy certain algebraic constraints. Hence, if $\frac{\partial F}{\partial \dot{x}}$ is singular, Eq. (2.1.2) is referred to as differential algebraic equation (DAE). In modeling, the formulation of pure ODE problems often requires the combination of conservation laws (mass and energy balance), constitutive equations (equations

© Atlantis Press and the author(s) 2016

N. Banagaaya et al., *Index-aware Model Order Reduction Methods*,
Atlantis Studies in Scientific Computing in Electromagnetics 2,
DOI 10.2991/978-94-6239-189-5_2

of state, pressure drops, heat transfer) and design constraints (desired operations). This means that there are some problems where not all the equations in a differential system involve derivatives, thus we can come up with a special case of differential algebraic equations which can be written as:

$$\dot{\boldsymbol{x}} = \boldsymbol{f}(t, \boldsymbol{x}, \boldsymbol{y}), \tag{2.1.3a}$$

$$\boldsymbol{0} = \boldsymbol{g}(t, \boldsymbol{x}, \boldsymbol{y}), \tag{2.1.3b}$$

where $\boldsymbol{x}$-differential variables, $\boldsymbol{y}$-algebraic variables and (2.1.3b) is a constraint equation. Equation (2.1.3) is a special type of DAE which is commonly called semi-explicit DAE.

2.2 Models for DAEs

According to [12], it is well known from modern control theory that two main mathematical representations for dynamical systems are the transfer matrix representation and the state space representation. The former describes only the input-output property of the system, while the latter gives further insight into the structural property of the system.

2.2.1 *State Space Representation*

State space representation was developed between the end of the 1950s and the beginning of the 1960s. It has the advantage that, not only it provides an efficient method for control system analysis and synthesis, but also offers a deeper understanding about the various properties of the systems. The state space models of the systems are obtained mainly using the so called state space variable method. To obtain a state model of a practical system, we need to choose some physical variables such as currents and voltages in an electrical network. Then, by the physical relationships among the variables or by some model identification techniques such as modified nodal analysis in network analysis, a set of equations can be established. Naturally, this set of equations are usually differential and/or algebraic equations, which form a mathematical model of the system. By properly defining a state vector $\boldsymbol{x}(t)$ and an input vector $\boldsymbol{u}(t)$, which are formed by the physical variables of the system, and an output vector $\boldsymbol{y}(t)$, whose elements are properly chosen measurable variables of the system, this set of equations can be arranged into two equations: one is the so called **state equation**, which can be written in the following implicit form:

$$\boldsymbol{f}(\dot{\boldsymbol{x}}(t), \boldsymbol{x}(t), \boldsymbol{u}(t), t) = 0, \tag{2.2.1}$$

and the second one is the **output equation**, or the observation equation, which can be in the general form of

$$\boldsymbol{g}(\boldsymbol{y}(t), \boldsymbol{x}(t), \boldsymbol{u}(t), t) = 0, \tag{2.2.2}$$

where $\boldsymbol{f}$ and $\boldsymbol{g}$ are vector functions of appropriate dimensions with respect to $\dot{\boldsymbol{x}}(t)$, $\boldsymbol{x}(t), \boldsymbol{y}(t), \boldsymbol{u}(t)$ and t. According to [12], Eqs. (2.2.1) and (2.2.2) give the state space representation for a general nonlinear dynamical system. We consider a special form of (2.2.1)–(2.2.2):

$$\mathbf{E}(t)\dot{\boldsymbol{x}}(t) = \mathbf{F}(\boldsymbol{x}(t), \boldsymbol{u}(t), t) \tag{2.2.3a}$$

$$\boldsymbol{y}(t) = \mathbf{K}(\boldsymbol{x}(t), \boldsymbol{u}(t), t), \tag{2.2.3b}$$

where $t \geq 0$ is the time variable, $\mathbf{F}$ and $\mathbf{K}$ are appropriate dimensional vector functions, $\boldsymbol{x}(t) \in \mathbb{R}^n$ is the state vector, $\boldsymbol{u}(t) \in \mathbb{R}^m$ is the control input vector, $\boldsymbol{y}(t) \in \mathbb{R}^p$ is the measured output vector. The matrix $\mathbf{E}(t)$ must be singular for some $t \geq 0$ for our case since we are considering DAEs. Equation (2.2.3) is the general form for the so called (nonlinear) DAEs. If we consider the case where $\mathbf{F}$ and $\mathbf{K}$ are linear functions of the vectors $\boldsymbol{x}(t)$ and $\boldsymbol{u}(t)$, the general nonlinear DAE (2.2.3) simplifies to the following form:

$$\mathbf{E}(t)\dot{\boldsymbol{x}}(t) = \mathbf{A}(t)\boldsymbol{x}(t) + \mathbf{B}(t)\boldsymbol{u}(t), \tag{2.2.4a}$$

$$\boldsymbol{y}(t) = \mathbf{C}^T(t)\boldsymbol{x}(t) + \mathbf{D}^T(t)\boldsymbol{u}(t), \tag{2.2.4b}$$

where $\mathbf{E}(t), \mathbf{A}(t) \in \mathbb{R}^{n\times n}$, $\mathbf{C}(t) \in \mathbb{R}^{n\times \ell}$, $\mathbf{D}(t) \in \mathbb{R}^{m\times \ell}$, $\mathbf{B}(t) \in \mathbb{R}^{n\times m}$ are matrix functions of time t, and they called the coefficients matrices of the system (2.2.4). Equation (2.2.4) describes the so called linear time-varying DAE. If the matrix coefficients are constant, i.e., time independent, the system (2.2.4) is called linear constant-coefficient DAE or linear time-invariant (LTI) DAE, which can be written as:

$$\mathbf{E}\dot{\boldsymbol{x}}(t) = \mathbf{A}\boldsymbol{x}(t) + \mathbf{B}\boldsymbol{u}(t), \tag{2.2.5a}$$

$$\boldsymbol{y}(t) = \mathbf{C}^T\boldsymbol{x}(t) + \mathbf{D}^T\boldsymbol{u}(t), \tag{2.2.5b}$$

where $\mathbf{E}, \mathbf{A} \in \mathbb{R}^{n\times n}$, $\mathbf{C} \in \mathbb{R}^{n\times \ell}$, $\mathbf{D} \in \mathbb{R}^{m\times \ell}$, $\mathbf{B} \in \mathbb{R}^{n\times m}$ are constant matrices. For the DAEs (2.2.5), there is also a concept of the **dynamical order**, which is defined as the rank of singular matrix $\mathbf{E}$. Equations (2.2.4) and (2.2.5) are the two basic classes of DAEs. From this point, we restrict ourselves on the DAEs of the form (2.2.5) unless stated otherwise.

2.2.2 *Transfer Matrix Representation*

In this section, we discuss the transfer matrix representation. This representation is derived from the state space representation using the Laplace transform. The transfer matrix representation is commonly used to validate reduced models in the model order reduction community and is commonly called the transfer function.

Definition 2.2.1 (*Laplace transform*) The Laplace transform of a function $f(t)$ in the time domain is the function $F(s)$ in the frequency domain and it is defined as: $\mathcal{L}\{f(t)\} = F(s) := \int_0^{\infty} e^{-st} f(t)\, \mathrm{d}t$, where $s = \sigma + j\omega \in \mathbb{C}$, with $\sigma, \omega \in \mathbb{R}$.

We shall restrict ourselves on the transfer matrix representation of the LTI DAEs (3.2.36) and also assume $s = j\omega$, i.e., $\sigma = 0$. Taking the Laplace transform of (3.2.36) and simplifying, we obtain

$$Y(s) = \left[\mathbf{C}^T(s\mathbf{E} - \mathbf{A})^{-1}\mathbf{B} + \mathbf{D}^T\right]\mathrm{U}(s) + \mathbf{C}^T(s\mathbf{E} - \mathbf{A})^{-1}\mathbf{E}\boldsymbol{x}(0), \tag{2.2.6}$$

where $\mathbf{U}(s)$ and $\mathbf{Y}(s)$ are the Laplace transforms of $\mathbf{u}(t)$ and $\mathbf{y}(t)$, respectively. The rational matrix-valued function

$$\mathbf{H}(s) = \mathbf{C}^T(s\mathbf{E} - \mathbf{A})^{-1}\mathbf{B} + \mathbf{D}^T \in \mathbb{R}^{\ell \times m}, \tag{2.2.7}$$

is called the transfer matrix representation of (2.2.5) or transfer function. Then, $\mathbf{H}(s)$ gives the relation between the Laplace transforms of the input $\mathbf{u}(t)$ and the output $\mathbf{y}(t)$. In other words, $\mathbf{H}(s)$ describes the input-output behavior of (2.2.5) in the frequency domain.

2.3 Linear Constant Coefficient DAEs

In this section, we discuss that analysis of the LTI DAEs. For simplicity, the coefficient matrix $\mathbf{D}$ in (2.2.5) is taken to be the zero matrix unless specified. Thus, (2.2.5) simplifies to:

$$\mathbf{E}\dot{\boldsymbol{x}}(t) = \mathbf{A}\boldsymbol{x}(t) + \mathbf{B}\boldsymbol{u}(t), \quad \boldsymbol{x}(0) = \boldsymbol{x}_0, \tag{2.3.1a}$$

$$\boldsymbol{y}(t) = \mathbf{C}^T\mathbf{x}(t), \tag{2.3.1b}$$

where $\mathbf{E} \in \mathbb{R}^{n\times n}$ is singular, $\mathbf{A} \in \mathbb{R}^{n\times n}$, $\mathbf{B} \in \mathbb{R}^{n\times m}$, $\mathbf{C} \in \mathbb{R}^{n\times \ell}$, and $\boldsymbol{u}(t) \in \mathbb{R}^m$ is the input vector, $\boldsymbol{y}(t) \in \mathbb{R}^\ell$ is the output vector and $\boldsymbol{x}_0 \in \mathbb{R}^n$ is the initial value.

2.3.1 Solvability of DAEs

Here, we are interested in the solutions of the homogenous system obtained by setting $\boldsymbol{u}(t) = 0$, then (2.3.1a) becomes

$$\mathbf{E}\dot{\boldsymbol{x}}(t) - \mathbf{A}\boldsymbol{x}(t) = \mathbf{0}. \qquad (2.3.2)$$

As for the case of ODEs we use the guess solution $\boldsymbol{x}(t) = \boldsymbol{x}_0 e^{\lambda t}$. Substituting the guess solution into (2.3.2) leads to $(\lambda\mathbf{E} - \mathbf{A})\boldsymbol{x}_0 e^{\lambda t} = \mathbf{0}$. We can see that (2.3.2) has a non-trivial solution provided $\lambda\mathbf{E} - \mathbf{A}$ is regular and $\boldsymbol{x}_0 \neq \mathbf{0}$ satisfies $(\lambda\mathbf{E} - \mathbf{A})\boldsymbol{x}_0 = \mathbf{0}$. λ and $\boldsymbol{x}_0$ are called the generalized eigenvalues and eigenvectors, respectively. Thus we say that (2.3.1a) is solvable provided the matrix pencil $\lambda\mathbf{E} - \mathbf{A}$ is regular. We note that $\lambda\mathbf{E} - \mathbf{A}$ can also be written as $(\mathbf{E}, \mathbf{A})$ which is called the matrix pencil or matrix pair. This leads to the following definition.

Definition 2.3.1 ([20]) A matrix pair $(\mathbf{E}, \mathbf{A})$ is called regular if the polynomial $\mathcal{P}(\lambda) = \det(\lambda\mathbf{E} - \mathbf{A})$ is not identically zero, it is called singular otherwise.

A pair $(\mathbf{E}, \mathbf{A})$ with nonsingular $\mathbf{E}$ is always regular, and its polynomial $\mathcal{P}(\lambda)$ is of degree n. In case of singular matrices $\mathbf{E}$, the polynomial is lower. According to [21], regularity of a matrix pair is closely related to the solution behavior of the corresponding DAE. In particular, regularity is necessary and sufficient for the property that for every sufficiently smooth inhomogeneity $\boldsymbol{u}$ the DAE is solvable and the solution is unique for every consistent initial value. This is well understood, if we consider the Weierstraß-Kronecker canonical form of a given DAE as discussed in Sect. 2.3.3.

Definition 2.3.2 ([34]) A pair $(\alpha, \beta) \in \mathbb{C}^2\backslash\{(0, 0)\}$ is said to be a generalized eigenvalue $\lambda = \frac{\alpha}{\beta}$ of the matrix pencil $\lambda\mathbf{E} - \mathbf{A}$ if $\det(\alpha\mathbf{E} - \beta\mathbf{A}) = 0$. If $\beta \neq 0$, then the pair (α, β) represents a finite eigenvalue $\lambda = \frac{\alpha}{\beta}$ of the matrix pencil $\lambda\mathbf{E} - \mathbf{A}$. But if $\beta = 0$, the pair $(\alpha, 0)$ represents an infinite eigenvalue of $\lambda\mathbf{E} - \mathbf{A}$. Clearly, the pencil $\lambda\mathbf{E} - \mathbf{A}$ has an eigenvalue at infinity if and only if the matrix $\mathbf{E}$ is singular.

The set of all finite eigenvalues of the matrix pencil $(\mathbf{E}, \mathbf{A})$ is denoted by $\sigma_f(\mathbf{E}, \mathbf{A})$ while the infinite spectrum of the matrix pencil $(\mathbf{E}, \mathbf{A})$ is denoted by $\sigma_\infty(\mathbf{E}, \mathbf{A})$. Thus, the set of all generalized eigenvalues (finite and infinite) of the matrix pencil (E;A) is called the spectrum of $(\mathbf{E}, \mathbf{A})$ and denoted by $\sigma(\mathbf{E}, \mathbf{A}) = \sigma_f(\mathbf{E}, \mathbf{A}) \cup \sigma_\infty(\mathbf{E}, \mathbf{A})$. We note that if $\mathbf{E}$ is nonsingular, then $\sigma(\mathbf{E}, \mathbf{A}) = \sigma_f(\mathbf{E}, \mathbf{A})$ which is equal to the spectrum of $\mathbf{E}^{-1}\mathbf{A}$. This also means that if $\mathbf{E}$ is nonsingular, the homogeneous equation (2.3.2) represents an implicit regular ODE and its fundamental solution system forms an n-dimensional subspace in $\mathcal{C}^1$. But what happens if $\mathbf{E}$ is singular, this is closely related to the notion of the regular matrix pencil $(\mathbf{E}, \mathbf{A})$ [20] as discussed in Sect. 2.3.3.

2.3.2 Stability of DAEs

According to [12], in practice a system should be stable, otherwise it may not work properly or may even be destroyed in practical use. As in the ODE case, when studying the stability of DAEs, we need to only consider the homogenous system (2.3.2).

Definition 2.3.3 ([34]) The DAE (2.3.1) is called asymptotically stable if $\lim_{t\to\infty} \boldsymbol{x}(t) = 0$ for all solutions $\boldsymbol{x}(t)$ of the homogenous system $\mathbf{E}\dot{\boldsymbol{x}}(t) = \mathbf{A}\boldsymbol{x}(t)$.

This leads us to the following theorem that collects equivalent conditions for system (2.3.1) to be asymptotically stable.

Theorem 2.3.1 *([34]) Consider a DAE (2.3.1) with regular matrix pencil $\lambda\mathbf{E} - \mathbf{A}$. The following statements are equivalent.*

1. *System (2.3.1) is asymptotically stable.*
2. *All finite eigenvalues of the matrix pencil $(\mathbf{E}, \mathbf{A})$ lie in the open left half complex plane, i.e., $\sigma(\mathbf{E}, \mathbf{A}) \subset \mathbb{C}^-$, where $\mathbb{C}^- = \{s \in \mathbb{C}, \mathrm{Re}(s) \leq 0\}$ represents the open left half complex plane.*

According to [34], the matrix pencil $\lambda\mathbf{E} - \mathbf{A}$ is called c-stable if it is regular and all the finite eigenvalues of $\lambda\mathbf{E} - \mathbf{A}$ have negative real part. We note that, in view of the above theorem, the infinite eigenvalues of the matrix pencil $(\mathbf{E}, \mathbf{A})$ have no effect on the stability of DAEs of the form (2.3.1), since the infinite eigenvalues of $\lambda\mathbf{E} - \mathbf{A}$ do not affect the behavior of the homogenous system at infinity [34]

2.3.3 Weierstraß-Kronecker Canonical Form

In this section, we present the Weierstraß-Kronecker canonical form. This is the most commonly used tool to understand the DAE structure of linear constant-coefficient DAEs [12, 21, 25]. Scaling (2.3.1) by nonsingular matrix $\mathbf{P} \in \mathbb{R}^{n\times n}$ and the state variable $\boldsymbol{x}$ according to $\boldsymbol{x} = \mathbf{Q}\tilde{\boldsymbol{x}}$ with a nonsingular matrix $\mathbf{Q} \in \mathbb{R}^{n\times n}$, we obtain

$$\tilde{\mathbf{E}}\dot{\tilde{\boldsymbol{x}}}(t) = \tilde{\mathbf{A}}\tilde{\boldsymbol{x}}(t) + \tilde{\mathbf{B}}\boldsymbol{u}(t), \qquad (2.3.3a)$$

$$\boldsymbol{y}(t) = \tilde{\mathbf{C}}^T\tilde{\boldsymbol{x}}(t), \qquad (2.3.3b)$$

where $\tilde{\mathbf{E}} = PEQ$, $\tilde{A} = PAQ$, $\tilde{B} = PB$ and $\tilde{\mathbf{C}} = \mathbf{Q}^{\mathrm{T}}\mathbf{C}$, which is again a DAE with constant coefficients. According to [21], the relation $\boldsymbol{x} = \mathbf{Q}\tilde{\boldsymbol{x}}$ gives a one-to-one correspondence between the corresponding solution sets. This means that we can consider the transformed problem (2.3.3) instead of (2.3.1) in order to understand the underlying structure of constant coefficients linear DAEs. It can easily be shown that this relation between systems (2.3.1) and (2.3.3) poses reflexivity, transitivity and symmetry, and is thus an equivalence relation.

Theorem 2.3.2 (Weierstraß-Kronecker canonical form [21, 29]) *Let* $(\mathbf{E}, \mathbf{A})$ *be a regular matrix pencil. Then, we have* $(E, A) \sim \left(\begin{pmatrix} \mathbf{I} & \mathbf{0} \\ \mathbf{0} & \mathbf{N} \end{pmatrix}, \begin{pmatrix} \mathbf{J} & \mathbf{0} \\ \mathbf{0} & \mathbf{I} \end{pmatrix}\right)$, *where* $\mathbf{J} \in \mathbb{R}^{k \times k}$ *for some nonnegative* $k \leq n$, *is a matrix in Jordan canonical form and* $\mathbf{N} \in \mathbb{R}^{(n-k)\times(n-k)}$ *is a nilpotent matrix with index* $\mu \leq n - k$ *also in Jordan canonical form. Moreover, it is allowed that one or the other block is not present.*

The equivalence symbol $\sim$ means that, the regular matrix pencil $\lambda\mathbf{E} - A$ can be transformed into $\lambda\tilde{\mathbf{E}} - \tilde{A}$, where $\tilde{\mathbf{E}} = \mathbf{PEQ} = \begin{pmatrix} \mathbf{I} & \mathbf{0} \\ \mathbf{0} & \mathbf{N} \end{pmatrix}$, $\tilde{\mathbf{A}} = \mathbf{PAQ} = \begin{pmatrix} \mathbf{J} & \mathbf{0} \\ \mathbf{0} & \mathbf{I} \end{pmatrix}$, by the use of suitable non-singular matrices $\mathbf{P}, \mathbf{Q} \in \mathbb{R}^{n \times n}$. There exists a $\mu \in \mathbb{N}$ such that $\mathbf{N}^{\mu-1} \neq 0$ but $\mathbf{N}^{\mu} = 0$, μ is known as $\mathbf{N}$'s index of nilpotency. μ is the index of the matrix pencil $\lambda\mathbf{E} - A$ and also the index of the differential algebraic system (2.3.1a). This index concept is commonly called the Kronecker index [12, 21, 29, 33, 34]. The numbers k and $n - k$ corresponds to the number of finite and infinite eigenvalues, respectively, of the spectrum of the matrix pencil $(\mathbf{E}, \mathbf{A})$. We note that it is possible to have $k = 0$, meaning $\tilde{\mathbf{E}} = \mathbf{N}$, $\tilde{\mathbf{A}} = \mathbf{I}$ this implies that the spectrum of the matrix pencil $\lambda\mathbf{E} - A$ has only infinite spectrum, i.e $\sigma(E, A) = \sigma_{\infty}(E, A)$ and, also $k = n$ which yields $\tilde{\mathbf{E}} = \mathbf{I}$, $\tilde{\mathbf{A}} = \mathbf{J}$, this implies that the spectrum of the matrix pencil $\sigma(E, A) = \sigma_f(E, A)$ has only finite spectrum. Assuming the matrix pencil $\lambda\mathbf{E} - A$ has both the finite and infinite spectrum, then the matrices $\tilde{\mathbf{B}}$ and $\tilde{\mathbf{C}}$ can be partitioned in blocks $\tilde{\mathbf{B}} = \begin{pmatrix} \tilde{\mathbf{B}}_1^{\mathrm{T}} & \tilde{\mathbf{B}}_2^{\mathrm{T}} \end{pmatrix}^{\mathrm{T}}$ and $\tilde{\mathbf{C}} = \begin{pmatrix} \tilde{\mathbf{C}}_1^{\mathrm{T}} & \tilde{\mathbf{C}}_2^{\mathrm{T}} \end{pmatrix}$, corresponding to the partitions of $\tilde{\mathbf{E}}$ and $\tilde{\mathbf{A}}$. Under the coordinate transformation $\tilde{\boldsymbol{x}} = \mathbf{Q}^{-1}\boldsymbol{x} = (\tilde{\boldsymbol{x}}_1^T(t), \tilde{\boldsymbol{x}}_2^T(t))^T$, system (2.3.3a) can be written as Weierstraß-Kronecker canonical form which leads to an equivalent decoupled system

$$\dot{\tilde{\boldsymbol{x}}}_1(t) = \mathbf{J}\tilde{\boldsymbol{x}}_1(t) + \tilde{\mathbf{B}}_1\boldsymbol{u}(t), \tag{2.3.4a}$$

$$\mathbf{N}\dot{\tilde{\boldsymbol{x}}}_2(t) = \tilde{\boldsymbol{x}}_2(t) + \tilde{\mathbf{B}}_2\boldsymbol{u}(t). \tag{2.3.4b}$$

We observe that (2.3.4a) represents a standard explicit ODE and without loss of generality the solution of (2.3.4b) can be written as

$$\tilde{\boldsymbol{x}}_2(t) = -\sum_{i=0}^{\mu-1} \mathbf{N}^i\tilde{\mathbf{B}}_2\boldsymbol{u}^{(i)}(t), \tag{2.3.5}$$

since $\mathbf{N}$ is a nilpotent matrix with index-μ, where $\boldsymbol{u}^{(i)}(t) = \frac{\mathrm{d}^i}{\mathrm{d}t}\boldsymbol{u}(t)$, provided $\mathbf{u}(t)$ is smooth enough, that is at least $\mu - 1$ times differentiable. Equation (2.3.5) shows the dependence of the solution $\boldsymbol{x}(t)$ of (2.3.1a) on the derivatives of the input function $\mathbf{u}(t)$. We can observe that the higher the index-μ, the more differentiations are involved. It is only in the index-1 case, where we have $\mathbf{N} = 0$, hence $\tilde{\boldsymbol{x}}_2(t) = \tilde{\mathbf{B}}_2\mathbf{u}(t)$, that no derivatives are involved. According to [20, 25], since numerical differentiations in these circumstances may cause considerable trouble numerically, it is very important to know the index of the DAE as well as details on the structure responsible for a higher index when modeling and simulating with DAEs in practice. From (2.3.4), we can observe that the number of finite eigenvalues, k and infinite eigenvalues, $n - k$ are equal to the number of differential and algebraic equations, respectively, in a given DAE. Thus, for the case of index-1 DAEs the number of differential equations is equal to the rank of the singular matrix $\mathbf{E}$. The solutions $\tilde{\boldsymbol{x}}_1$ and $\tilde{\boldsymbol{x}}_2$ which corresponds to the differential and algebraic part are commonly known as the slow and fast solutions, respectively, see [12, 21, 29, 33, 34]. It can be easily proved that the differential part of (2.3.4) inherits the stability of DAEs of the form (2.3.1), that is, $\sigma_f(\mathbf{E}, \mathbf{A}) = \sigma(\mathbf{J})$.

2.3.3.1 Index Concept of DAEs

An index of a DAE is commonly defined as the measure of the difficulties arising in the theoretical and numerical treatment of a given DAE. According to [21], the motivation to introduce an index is to classify different types of DAEs with respect to the difficulty to solve them analytically as well as numerically. Sometimes the index of a DAE is defined as a measure of how much the DAE deviates from an ODE. In the previous Section, we have defined the index μ as the nilpotency index of a nilpotent matrix $\mathbf{N}$. This index is also known as the Kronecker index of a DAE. Many other index concepts have been introduced, see [8, 29, 32], but in this book we shall restrict ourselves to only three, i.e., Kronecker index, differentiation index and tractability index. We note that all these index concepts coincide for the case of linear DAEs with constant matrices. If we differentiate (2.3.12b) with respect to t we obtain: $\dot{\tilde{\boldsymbol{x}}}_2 = -\sum_{i=0}^{\mu-1} \mathbf{N}^i \tilde{\mathbf{B}}_2 \boldsymbol{u}^{(i+1)}$. This means that exactly μ differentiations transform (3.2.48) into a system of explicit ordinary differential equations. Hence, the Kronecker and the differentiation index coincide for LTI DAEs. This type of index is called the differentiation index and it is defined as in Definition 2.3.4. According to [21], the differentiation index was introduced to determine how far the DAE is away from an ODE, for which the analysis and numerical techniques are well established.

Definition 2.3.4 ([32]) The nonlinear DAE, $\boldsymbol{F}(t, \boldsymbol{x}, \dot{\boldsymbol{x}}) = \mathbf{0}$, has differentiation index μ , if μ is the minimal number of differentiations

$$\boldsymbol{F}(t, \boldsymbol{x}, \dot{\boldsymbol{x}}) = \mathbf{0}, \ \frac{d}{dt}\big(\boldsymbol{F}(t, \boldsymbol{x}, \dot{\boldsymbol{x}})\big) = \mathbf{0}, \ldots, \ \frac{d^{\mu}}{dt}\big(\boldsymbol{F}(t, \boldsymbol{x}, \dot{\boldsymbol{x}})\big) = \mathbf{0}, \tag{2.3.6}$$

such that the Eq. (2.3.6) make it possible to extract an explicit ordinary differentiation system $\dot{\boldsymbol{x}} = \boldsymbol{f}(t, \boldsymbol{x})$ using only algebraic manipulations.

This can be illustrated as follows. Consider a semi-explicit DAE of the form below:

$$\dot{\boldsymbol{x}} = \boldsymbol{f}(\boldsymbol{x}, \boldsymbol{y}), \tag{2.3.7a}$$

$$\mathbf{0} = \mathbf{g}(\boldsymbol{x}, \boldsymbol{y}). \tag{2.3.7b}$$

Using chain rule on the constraint equation, we find:

$$\dot{x} = f(x, y), \tag{2.3.8a}$$
$$\dot{y} = -g_y(x, y)^{-1}\left(g_x(x, y)f(x, y)\right). \tag{2.3.8b}$$

Thus, the DAE has a differentiation index $\mu = 1$ provided $\det(g_y) \neq 0$. In the example below, we compare the differential index and the Kronecker index.

Example 2.3.1 Consider a DAE system of the form (2.3.1) with system matrices,

$$\mathbf{E} = \begin{pmatrix} 1\,0\,0\,0 \\ 0\,0\,1\,0 \\ 0\,0\,0\,0 \\ 0\,0\,0\,0 \end{pmatrix}, \mathbf{A} = \begin{pmatrix} 0 & 1\,0\,0 \\ 1 & 0\,0\,0 \\ -1 & 0\,0\,1 \\ 0 & 1\,1\,1 \end{pmatrix}, \mathbf{B} = \begin{pmatrix} 0 \\ 0 \\ 0 \\ -1 \end{pmatrix}, \quad \text{and} \quad \mathbf{C} = \begin{pmatrix} 0 \\ 0 \\ 1 \\ 0 \end{pmatrix}. \tag{2.3.9}$$

The matrix pencil $(\mathbf{E}, \mathbf{A})$ is regular since $\det(\lambda\mathbf{E} - \mathbf{A}) = \lambda^2 + \lambda + 1 \neq 0$. Using Theorem (2.3.2) we can choose non-singular matrices, $\mathbf{Q} = \begin{pmatrix} 1\,0 & 1 & -1 \\ 0\,1 & 0 & 0 \\ 0\,0 & -1 & 1 \\ 0\,0 & 1 & 0 \end{pmatrix}$ and $\mathbf{P} = \begin{pmatrix} 1 & 0 & 0\,0 \\ -1 & -1 & 1\,0 \\ 0 & 1 & 0\,0 \\ 1 & 0 & 0\,1 \end{pmatrix}$ such that $(\mathbf{E}, \mathbf{A}) \sim \left(\begin{pmatrix} \mathbf{I}\,\mathbf{0} \\ \mathbf{0}\,\mathbf{N} \end{pmatrix}, \begin{pmatrix} \mathbf{J}\,\mathbf{0} \\ \mathbf{0}\,\mathbf{I} \end{pmatrix}\right)$, where $\mathbf{J} = \begin{pmatrix} -1 & -1 \\ 1 & 0 \end{pmatrix}$ and $\mathbf{N} = \begin{pmatrix} 0\,0 \\ 0\,0 \end{pmatrix}$. Thus this DAE system is of Kronecker index 1, since $\mathbf{N}$ is of nilpotent index 1. Next, we compute the differential index. We need first to rewrite system (2.3.9) in the the semi-explicit form (2.3.7) where $x = \begin{pmatrix} x_1 \\ x_3 \end{pmatrix}$, $\mathbf{y} = \begin{pmatrix} x_2 \\ x_4 \end{pmatrix}$, $f(x, y) = \begin{pmatrix} x_2 \\ x_1 \end{pmatrix}$ and $g(x, y) = \begin{pmatrix} -x_1 + x_4 \\ x_2 + x_3 + x_4 \end{pmatrix}$. Then $g_y = \begin{pmatrix} 0\,1 \\ 1\,1 \end{pmatrix}$, with $\det(g_y) = -1 \neq 0$. Thus the DAE system has differentiation index 1. Hence the Kronecker index and differentiation index coincide, that is, $\mu = \gamma = 1$.

2.3.3.2 Consistent Initial Condition of DAEs

From system (2.3.4), we observe that (2.3.4a) is a linear differential equation which can easily be solved when any initial condition $\tilde{x}_1(0)$ is applied and its analytic solution can be written as:

$$\tilde{x}_1(t) = e^{t\mathbf{J}}\tilde{x}_1(0) + e^{t\mathbf{J}} \int_0^t e^{-\tau\mathbf{J}}\tilde{\mathbf{B}}_1 u(\tau)\, d\tau. \tag{2.3.10}$$

We observe that the solution (2.3.10) of (2.3.12a) is always unique for any choice of the initial value $\tilde{\boldsymbol{x}}_1(0)$ while the initial value of (2.3.12b) has to satisfy the hidden constraint,

$$\tilde{\boldsymbol{x}}_2(0) = -\sum_{i=0}^{\mu-1} \mathbf{N}^i \tilde{\mathbf{B}}_2 \boldsymbol{u}^{(i)}(0). \tag{2.3.11}$$

We can observe that, we have no enough freedom to arbitrary choose the initial values $\tilde{\boldsymbol{x}}_2(0)$. For example, if $\mu = 1$, we have to choose the initial value such that $\tilde{\boldsymbol{x}}_2(0) = -\tilde{\mathbf{B}}_2\boldsymbol{u}(0)$. For $\mu > 1$, Eq. (2.3.5) is a differentiation problem, thus the initial value $\tilde{\boldsymbol{x}}_2(0)$ is fixed, and the input function $\boldsymbol{u}(t)$ has to be at least $\mu - 1$ times differentiable, i.e., $\boldsymbol{u}(t) \in \mathcal{C}^{\mu-1}$. Hence, initial value problems for (2.3.1) lead to unique classical solutions if the initial value $\boldsymbol{x}(0) = \boldsymbol{x}_0$ is consistent, that is, $\boldsymbol{x}(0) = \mathbf{Q}\tilde{\boldsymbol{x}}(0) = \mathbf{Q}\left[\tilde{\boldsymbol{x}}_1(0)^T \; \tilde{\boldsymbol{x}}_2(0)^T\right]^T$, where $\tilde{\boldsymbol{x}}_1(0)$ is a free parameter while $\tilde{\boldsymbol{x}}_2(0)$ has to satisfy (2.3.11). Thus, $\boldsymbol{x}(0)$ must be a consistent initial condition of a given DAE. We note that, if the initial condition $\boldsymbol{x}_0$ is inconsistent or the input function $\boldsymbol{u}(t)$ is not sufficiently smooth, then the solution of the DAEs may have impulsive modes, see [12, 33].

2.3.4 Transfer Matrix Representation of the Kronecker Form

Using (2.3.4) and the decomposed output equation, control problem (2.3.1) can also be written in equivalent form

$$\dot{\tilde{\boldsymbol{x}}}_1(t) = \mathbf{J}\tilde{\boldsymbol{x}}_1(t) + \tilde{\mathbf{B}}_1\boldsymbol{u}(t), \tag{2.3.12a}$$

$$\tilde{\boldsymbol{x}}_2(t) = -\sum_{i=0}^{\mu-1} \mathbf{N}^i \tilde{\mathbf{B}}_2 \boldsymbol{u}^{(i)}(t), \tag{2.3.12b}$$

$$\boldsymbol{y}(t) = \tilde{\mathbf{C}}_1^T\tilde{\boldsymbol{x}}_1(t) + \tilde{\mathbf{C}}_2^T\tilde{\boldsymbol{x}}_2(t). \tag{2.3.12c}$$

Taking the Laplace transform of (2.3.12), using the fact that

$$\mathcal{L}\{f^{(n)}(t)\} = s^n\mathcal{L}\{f(t)\} - \sum_{k=1}^{n} s^{k-1} f^{(n-k)}(0),$$

and setting $\tilde{\boldsymbol{x}}_1(0) = 0$ we obtain:

$$\mathbf{Y}(s) = \tilde{\mathbf{H}}(s)\mathbf{U}(s) + \tilde{\mathbf{C}}_2^T \sum_{i=0}^{\mu-1} \mathbf{N}^i \tilde{\mathbf{B}}_2 \sum_{k=1}^{i} s^{k-1}\mathbf{u}^{(i-k)}(0), \tag{2.3.13}$$

where $\tilde{\mathbf{H}}(s) = \mathbf{H}_1(s) + \mathbf{H}_2(s)$, $\mathbf{H}_1(s) = \tilde{\mathbf{C}}_1^T(s\mathrm{I} - \mathbf{J})^{-1}\tilde{\mathbf{B}}_1$ and $\mathbf{H}_2(s) = -\tilde{\mathbf{C}}_2^T\sum_{i=0}^{\mu-1} \mathbf{N}^i\tilde{\mathbf{B}}_2 s^i$. It can be easily proved that $\tilde{\mathbf{H}}(s)$ coincides with the conventional definition of the transfer function $\mathbf{H}(s) = \mathbf{C}^{\mathrm{T}}(s\mathbf{E} - \mathbf{A})^{-1}\mathbf{B}$, since (2.3.1) and (2.3.12) are equivalent, i.e., $\mathbf{H}(s) = \tilde{\mathbf{H}}(s)$. If let $\mathcal{P}(s) = \tilde{\mathbf{C}}_2^T\sum_{i=0}^{\mu-1}\mathbf{N}^i\tilde{\mathbf{B}}_2\sum_{k=1}^{i} s^{k-1}\mathbf{u}^{(i-k)}(0)$, Eq. (2.3.13) can be decomposed as $\mathbf{Y}(s) = \mathbf{Y}_1(s) + \mathbf{Y}_2(s)$, where $\mathbf{Y}_1(s) = \mathbf{H}_1(s)\mathbf{U}(s)$ and $\mathbf{Y}_2(s) = \mathbf{H}_2(s)\mathbf{U}(s) + \mathcal{P}(s)$ represent the input-output function of the differential and algebraic parts, respectively in the frequency domain. We note that the $\mathbf{H}_1(s)$ and $\mathbf{H}_2(s)$ are commonly called the strictly proper and the polynomial parts of $\mathbf{H}(s)$, respectively, see [33]. We observe that the input-output relation of the differential part is given by $\mathbf{Y}_1(s) = \mathbf{H}_1(s)\mathbf{U}(s)$, where $\mathbf{H}_1(s)$ is its transfer matrix representation and it is independent of the index-μ of the DAE (3.2.48) while input-output relation of the algebraic part is given by $\mathbf{Y}_2(s) = \mathbf{H}_2(s)\mathbf{U}(s) + \mathcal{P}(s)$ which depends on the index-μ of the DAE (3.2.48). We note that $\mathcal{P}(s) = 0$ always for the case of index-1 DAEs. According to [24, 29], transforming Eq. (2.3.1) into a Kronecker canonical form is just in theory, but practical implementation may be difficult or impossible. This is due to the fact that computing the Kronecker canonical form in finite precision arithmetic is, in general, an ill-conditioned problem in the sense that small changes in the data may extremely change the Kronecker canonical form [33]. Proper formulations and projector methods attempt to overcome these drawbacks, allowing additionally for an extension of the results to the time varying context [29]. These techniques provide an index characterization in terms of the original problem description. This motivated us to use the projector and matrix chains approach in order to decompose LTI DAEs into differential and algebraic parts as introduced by März [25]. This will be discussed in Chap. 3.

Chapter 3
Decoupling of Linear Constant DAEs

In this chapter, we discuss how to decouple DAEs using matrix, projector and basis chains. This approach is based on the projector and matrix chains introduced in [25].

3.1 Decoupling of DAEs Using Matrix and Projector Chains

In order to decouple DAEs using matrix and projector chains, we need first to discuss their construction. This can be done as follows. A square matrix $\mathbf{Q}$ is called projector if and only if $\mathbf{Q}^2 = \mathbf{Q}$. A projector $\mathbf{Q}$ is called projector onto a subspace $S \subset \mathbb{R}^n$ if $\operatorname{Im} \mathbf{Q} = S$. It is called projector along a subspace $S \subset \mathbb{R}^n$ if $\operatorname{Ker} \mathbf{Q} = S$.

Definition 3.1.1 (*Tractability index* [25]) Let $(\mathbf{E}, \mathbf{A})$ be a regular matrix pair. We define a matrix and projector chain by setting $\mathbf{E}_0 := \mathbf{E}$ and $\mathbf{A}_0 := \mathbf{A}$, then

$$\mathbf{E}_{j+1} := \mathbf{E}_j - \mathbf{A}_j\mathbf{Q}_j, \quad \mathbf{A}_{j+1} := \mathbf{A}_j\mathbf{P}_j, \text{ for} j \geq 0, \tag{3.1.1}$$

where $\mathbf{Q}_j$ are projectors onto $\operatorname{Ker} \mathbf{E}_j$ and $\mathbf{P}_j = \mathrm{I} - \mathbf{Q}_j$. There exists an index μ such that $\mathbf{E}_\mu$ is non-singular and all $\mathbf{E}_j$ are singular for all $0 \leq j < \mu - 1$. This type of index is called the *tractability index* and we say that matrix pair $(\mathbf{E}, \mathbf{A})$ has *tractability* index μ and it is denoted by $\operatorname{ind}(\mathbf{E}, \mathbf{A}) = \mu$.

The matrix and projector chain from the above definition, can only be used to decouple index-1 DAEs. For higher index we need to introduce some extra constraints. This lead us to the following theorem.

© Atlantis Press and the author(s) 2016

N. Banagaaya et al., *Index-aware Model Order Reduction Methods*,
Atlantis Studies in Scientific Computing in Electromagnetics 2,
DOI 10.2991/978-94-6239-189-5_3

Theorem 3.1.1 [25] *Given a regular index μ DAE (3.2.48).*

(i) The $\mathbf{E}_0, \dots, \mathbf{E}_{\mu-1}$ *are singular but* $\mathbf{E}_\mu$ *is nonsingular.*
(ii) $\dim \operatorname{Ker} \mathbf{E}_{j+1} = \dim S_j \cap \operatorname{Ker} \mathbf{E}_j$, *where* $S_j = \{z \in \mathbb{R}^n : \mathbf{A}_i z = \operatorname{Ker} \mathbf{E}_j\}$, $j \geq 0$.
(iii) The projectors $\mathbf{Q}_0, \dots, \mathbf{Q}_{\mu-1}$ *may be chosen such that*

$$\mathbf{Q}_j \mathbf{Q}_i = 0, \quad \textit{for } j > i. \tag{3.1.2}$$

We note that projectors satisfying the property (3.1.2) in Theorem 3.1.1 are commonly known as the admissible projectors. Moreover, for admissible projectors $\mathbf{Q}_0, \dots, \mathbf{Q}_{\mu-1}$, we have $\operatorname{Ker}(\mathbf{P}_0\mathbf{P}_1 \dots \mathbf{P}_j) = \operatorname{Im} \mathbf{Q}_0 \oplus \dots \oplus \operatorname{Im} \mathbf{Q}_j$. It can also be verified that for admissible projectors the following absorption properties holds, $\mathbf{P}_j \mathbf{Q}_i = \mathbf{Q}_i$, $\mathbf{Q}_j \mathbf{P}_i = \mathbf{Q}_j$, $\forall j > i$, which implies, $\mathbf{P}_j \mathbf{P}_{j-1} \dots \mathbf{P}_0 = \mathrm{I} - \sum_{i=0}^{j} \mathbf{Q}_i$, $\forall j > 0$. Without loss of generality, using matrix and admissible projector chains, (3.2.48) can be written as

$$\mathbf{E}_\mu \Big[\mathbf{P}_{\mu-1} \dots \mathbf{P}_0 \dot{\boldsymbol{x}} + \mathbf{Q}_0 \boldsymbol{x} + \dots + \mathbf{Q}_{\mu-1} \boldsymbol{x}\Big] = \mathbf{A}_\mu \boldsymbol{x} + \mathbf{B}\boldsymbol{u}. \tag{3.1.3}$$

Next, we need to decouple (3.1.3) into differential and algebraic parts. This is done as follow. According to [25], since $\mathbf{E}_\mu$ is nonsingular, then (3.1.3) can be written as,

$$\mathbf{P}_{\mu-1} \dots \mathbf{P}_0 \dot{\boldsymbol{x}} + \mathbf{Q}_0 \boldsymbol{x} + \dots + \mathbf{Q}_{\mu-1} \boldsymbol{x} = \mathbf{E}_\mu^{-1} \Big[\mathbf{A}_\mu \boldsymbol{x} + \mathbf{B}\boldsymbol{u}\Big]. \tag{3.1.4}$$

In order to decouple (3.1.4), we need to decompose the identity matrix into two ways:

$$\mathrm{I} = \mathbf{P}_0 + \mathbf{Q}_0 = \Pi_{\mu-1} + \Pi_{\mu-2}\mathbf{Q}_{\mu-1} + \dots + \Pi_0 \mathbf{Q}_1 + \mathbf{Q}_0, \tag{3.1.5}$$
$$\mathrm{I} = \mathbf{P}_{\mu-1} + \mathbf{Q}_{\mu-1} = \Pi_0^* + \mathbf{Q}_0 \Pi_1^* + \dots + \mathbf{Q}_{\mu-2} \Pi_{\mu-1}^* + \mathbf{Q}_{\mu-1}, \tag{3.1.6}$$

where $\Pi_j := \mathbf{P}_0 \mathbf{P}_1 \dots \mathbf{P}_j$, $\Pi_j^* = \mathbf{P}_j \mathbf{P}_{j+1} \dots \mathbf{P}_{\mu-1}$, $j = 0, 1, \dots, \mu - 1$. We note that both of the above decompositions of identity matrix are made up of mutually orthogonal projectors since we are using admissible projectors. We also note that if we do not use admissible projectors and we just use the projector chains in Definition 3.1.1, the above decompositions do not hold. Hence we cannot have a decoupling of (3.1.4) with index $\mu > 1$. We can now use these two decompositions to decompose higher index DAE into differential and algebraic equations. This can be done as follows: Decomposition (3.1.5) is used to define the differential and algebraic components:

$$\boldsymbol{x}_P := \Pi_{\mu-1} \boldsymbol{x}, \quad \boldsymbol{x}_{Q,0} := \mathbf{Q}_0 \boldsymbol{x}, \quad \boldsymbol{x}_{Q,i} := \Pi_{i-1} \mathbf{Q}_i \boldsymbol{x}, \quad i = 1, \dots \mu - 1. \tag{3.1.7}$$

The second decomposition (3.1.6) is used to derive the differential and algebraic equations. Without loss of generality, if the DAE (2.3.1) is of tractability index μ and the spectrum of its matrix pencil has at least one finite eigenvalue, then its equivalent decoupled system is given by

$$\boldsymbol{x}_P' = \mathbf{A}_P\boldsymbol{x}_P + \mathbf{B}_P u, \quad \boldsymbol{x}_P(0) = \Pi_{\mu-1}x(0), \tag{3.1.8a}$$

$$\boldsymbol{x}_{Q,\mu-1} = \mathbf{A}_{Q,\mu-1}\boldsymbol{x}_P + \mathbf{B}_{Q,\mu-1}u, \tag{3.1.8b}$$

$$\boldsymbol{x}_{Q,i} = \mathbf{A}_{Q,i}\boldsymbol{x}_P + \mathbf{B}_{Q,i}u + \sum_{j=i+1}^{\mu-1}\mathbf{A}_{\mathbf{Q}_{i,j}}\boldsymbol{x}_{Q,j}', \; i = \mu-2,\ldots 0, \tag{3.1.8c}$$

$$\mathbf{y} = \mathbf{C}^T\boldsymbol{x}_P + \mathbf{C}^T\sum_{i=0}^{\mu-1}\boldsymbol{x}_{Q,i}, \tag{3.1.8d}$$

where $\mathbf{A}_P := \Pi_0^*\mathbf{E}_\mu^{-1}\mathbf{A}_\mu,\;\; \mathbf{B}_P := \Pi_0^*\mathbf{E}_\mu^{-1}\mathbf{B},\;\; \mathbf{A}_{Q,\mu-1} := \Pi_{\mu-2}\mathbf{Q}_{\mu-1}\mathbf{E}_\mu^{-1}\mathbf{A}_\mu,$ $\mathbf{B}_{Q,\mu-1} := \Pi_{\mu-2}\mathbf{Q}_{\mu-1}\mathbf{E}_\mu^{-1}\mathbf{B},\;\; \mathbf{A}_{Q,0} := \mathbf{Q}_0\Pi_1^*\mathbf{E}_\mu^{-1}\mathbf{A}_\mu,\;\; \mathbf{B}_{Q,0} := \mathbf{Q}_0\Pi_1^*\mathbf{E}_\mu^{-1}\mathbf{B},$
$\mathbf{A}_{\mathbf{Q}_{i,j}} := \Pi_{i-1}\mathbf{Q}_{i,j},\;\; \mathbf{Q}_{i,j} = \begin{cases}\mathbf{Q}_i\mathbf{Q}_{i+1}, & j = i+1,\\ \mathbf{Q}_i\mathbf{P}_{i+1}\ldots\mathbf{P}_{j-1}\mathbf{Q}_j, & j > i+1.\end{cases}$
System (3.1.8) can be solved in the following way: first, the differential part $\boldsymbol{x}_P$ is computed from the purely differential equation (3.1.8a); then the algebraic parts are computed, starting from the last one, $\boldsymbol{x}_{Q,\mu-1}$, given by (3.1.8c), and substituting the computed values in the last but one equation for $\boldsymbol{x}_{Q,\mu-2}$, given by (3.1.8c) for $i = \mu-2$, and so on, up to the first equation for $\boldsymbol{x}_{Q,0}$. Finally, the desired output solutions can be obtained through (3.1.8d). From (3.1.8), we can observe that the system is still not completely decoupled, in fact there is a one way coupling between the differential and algebraic parts. However, it is much easier to solve (3.1.8) and it can be beneficial if we can obtain a completely decoupled system. Fortunately, they are already existing ways to achieve this. In [25] the so called canonical projectors were introduced. Canonical projectors are admissible projectors that satisfy the following theorem.

Theorem 3.1.2 [25] *Let* $(\mathbf{E}, \mathbf{A})$ *be a regular index-*μ *matrix pencil. Then there are projectors* $\mathbf{Q}_0, \ldots, \mathbf{Q}_{\mu-1}$ *such that*

$$\mathbf{Q}_j = -\mathbf{Q}_j\Pi_{j+1}^*\mathbf{E}_\mu^{-1}\mathbf{A}_j, \;\; j = 0,\ldots \mu-2,$$
$$\mathbf{Q}_{\mu-1} = -\mathbf{Q}_{\mu-1}\mathbf{E}_\mu^{-1}\mathbf{A}_{\mu-1}.$$

The proof can be found in [25]. Hence, if we choose canonical projectors in advance, then the coupling terms in (3.1.8) disappear. Thus, (3.1.8) can be simplified to

$$\dot{\boldsymbol{x}}_P = \mathbf{A}_P\boldsymbol{x}_P + \mathbf{B}_P u, \quad \boldsymbol{x}_P(0) = \Pi_{\mu-1}x(0), \tag{3.1.9a}$$

$$\boldsymbol{x}_{Q,\mu-1} = \mathbf{B}_{Q,\mu-1}u, \tag{3.1.9b}$$

$$x_{Q,i} = \mathbf{B}_{Q,i}u + \sum_{j=i+1}^{\mu-1} \mathbf{A}_{\mathbf{Q}_{i,j}} x'_{Q,j}, \; i = \mu - 2, \ldots 0, \quad i = \mu - 2, \ldots 0, \tag{3.1.9c}$$

But, one may wonder whether the construction of canonical projectors is feasible in practice. Fortunately, they do exist in practice and their construction is well discussed in [27, 37] which we briefly discuss in the next section.

3.1.1 Construction of Canonical Projectors

In the following recursive construction of matrix and canonical projector chains, we denote by $\mathbf{E}_j^{(i)}$, $\mathbf{A}_j^{(i)}$, $\mathbf{Q}_j^{(i)}$, $\mathbf{P}_j^{(i)}$ the ith iterate of $\mathbf{E}_j$, $\mathbf{A}_j$, $\mathbf{Q}_j$, $\mathbf{P}_j$ in the recursive construction. Here, we start the iteration with the matrices and projectors $\mathbf{E}_j^{(0)}$, $\mathbf{A}_j^{(0)}$, $\mathbf{Q}_j^{(0)}$, $\mathbf{P}_j^{(0)}$ constructed using Definition 3.1.1, and $\mathbf{Q}_j$ are admissible projectors. The construction of admissible projectors from Definition 3.1.1 is well discussed in [27]. Here, we assume that projector $\mathbf{Q}_j$ are already admissible. Then from admissible projectors, we can construct canonical projectors using Theorem 3.1.2 and the formula is given by

$$\mathbf{Q}_{\mu-1}^{(k)} = -\mathbf{Q}_{\mu-1}^{(k-1)}(\mathbf{E}_{\mu}^{(k-1)})^{-1}\mathbf{A}_{\mu-1}^{(k-1)}, \tag{3.1.10a}$$

$$\mathbf{Q}_{j}^{(k)} = -\mathbf{Q}_{j}^{(k-1)}\Pi_{j+1}^{*(k-1)}(\mathbf{E}_{\mu}^{(k-1)})^{-1}\mathbf{A}_{j}^{(k-1)}, \quad j = 0, \ldots \mu - 2, \tag{3.1.10b}$$

where $k = 1, \ldots, \mu$ is the number of iterations or updates to converge to canonical projectors. We note that $(\mathbf{E}_{\mu}^{(k)})^{-1}$ is to be computed explicitly, which in practice makes the canonical projector construction computationally expensive, see [37]. According to [37], the optimal number of updates to converge to canonical projectors is μ. However, more iterations can be done but the canonical projector chains will just repeat themselves. Below, we give some examples of construction of canonical projectors for the case of index-1 and index-2 DAEs.

(i) **Index-1 matrix pencil** $(\mathbf{E}, \mathbf{A})$. Here, the canonical projector construction takes $k = 1$ update or iterate. This is given by
$\mathbf{Q}_0^{(1)} = -\mathbf{Q}_0^{(0)}\mathbf{E}_1^{(0)^{-1}}\mathbf{A}_0^{(0)}$ and $\mathbf{P}_0^{(1)} = \mathrm{I} - \mathbf{Q}_0^{(1)}$. Hence the desired matrix and canonical projector chains which can lead to a completely decoupled index-1 DAEs is given by $\mathbf{Q}_0 = \mathbf{Q}_0^{(1)}$.

(ii) **Index-2 matrix pencil** $(\mathbf{E}, \mathbf{A})$. Here, the canonical projector construction takes $k = 2$ iterates as follows:

(i) $\mathbf{Q}_1^{(1)} = -\mathbf{Q}_1^{(0)}(\mathbf{E}_2^{(0)})^{-1}\mathbf{A}_1^{(0)}, \quad \mathbf{Q}_0^{(1)} = -\mathbf{Q}_0^{(0)}\mathbf{P}_1^{(0)}(\mathbf{E}_2^{(0)})^{-1}\mathbf{A}_0^{(0)},$

(ii) $\mathbf{E}_2^{(1)} = \mathbf{E}_1^{(1)} - \mathbf{A}_1^{(1)}\mathbf{Q}_1^{(1)}, \quad \mathbf{A}_1^{(1)} = \mathbf{A}_0^{(0)}\mathbf{P}_0^{(1)},$

$\mathbf{Q}_1^{(2)} = -\mathbf{Q}_1^{(1)}(\mathbf{E}_2^{(1)})^{-1}\mathbf{A}_1^{(1)}, \quad \mathbf{Q}_0^{(2)} = -\mathbf{Q}_0^{(1)}\mathbf{P}_1^{(1)}(\mathbf{E}_2^{(1)})^{-1}\mathbf{A}_0^{(1)}.$

Hence the desired matrix and canonical projector chains which can lead to a completely decoupled index-2 DAEs are given by $\mathbf{Q}_0 = \mathbf{Q}_0^{(2)}$ and $\mathbf{Q}_1 = \mathbf{Q}_1^{(2)}$.

We note that canonical projectors for higher index DAEs can be constructed from admissible projectors using the formula (3.1.10), see [37] for more details.

3.2 Decoupling of DAEs Using Matrix, Projector and Basis Chains

We observe that the total dimension of (3.1.9) or (3.1.8) is $n(1+\mu)$ while the dimension of the DAE (3.2.48). This implies that using matrix and projector chains increases the dimension of the DAE system. In [2, 4], it is proposed a way of overcoming this problem by using the basis chains of the corresponding projector chains and their products. This implies that the decoupling of DAEs is now done by using matrix, projector and basis chains as follows. We consider the decomposition of the identity matrix (3.1.5). We remove the redundancy in (3.1.9) or (3.1.8) by using basis column matrices for the special projectors and their products. From (3.1.5), we have:

$$\mathbf{I}_n = \mathbf{Q}_0 + \sum_{i=1}^{\mu-1} \Pi_{i-1}\mathbf{Q}_i + \Pi_{\mu-1}. \tag{3.2.1}$$

We have to note that higher index DAEs have a possibility of having a purely algebraic system depending to the nature of the spectrum of the matrix pencil $(\mathbf{E}, \mathbf{A})$. This implies the projector product $\Pi_{\mu-1}$ can vanish to zero depending on the matrix pencil of the DAE (3.2.48). Thus in this section, we consider two cases of compact decomposition of the DAE (3.2.48) depending on the spectrum of the matrix pencil $(\mathbf{E}, \mathbf{A})$.

3.2.1 Matrix Pencil (E, A) *with Atleast One Finite Eigenvalue*

Here, we assume that the spectrum of the matrix pencil of (3.2.48) has at least one finite eigenvalue, this implies $\Pi_{\mu-1} \neq 0$ in (3.2.1). If, we let $\boldsymbol{q}_0 \in \operatorname{Ker} \mathbf{E}_0$ and $\boldsymbol{p}_0 \in \operatorname{Im} \mathbf{P}_0$, then $\mathbf{Q}_0\boldsymbol{q}_0 = \boldsymbol{q}_0$, $\mathbf{P}_0\boldsymbol{q}_0 = \boldsymbol{q}_0$, $\mathbf{P}_0\boldsymbol{p}_0 = \boldsymbol{p}_0$, $\mathbf{Q}_0\boldsymbol{p}_0 = \boldsymbol{q}_0$. Let $k_0 = \dim(\operatorname{Ker} \mathbf{E}_0)$, $n_0 = n - k_0$. Then, we can build an orthonormal basis matrix $(\boldsymbol{p}_0, \boldsymbol{q}_0) = (\boldsymbol{p}_{0,1}, \ldots, \boldsymbol{p}_{0,n_0}, \boldsymbol{q}_{0,1}, \ldots, \boldsymbol{q}_{0,k_0}) \in \mathbb{R}^{n\times n}$. Since $(\boldsymbol{p}_0, \boldsymbol{q}_0)$ is a basis matrix, it is invertible, and its inverse is denoted by $(\boldsymbol{p}_0, \boldsymbol{q}_0)^{-1} = (\boldsymbol{p}_0^{*T}, \boldsymbol{q}_0^{*T})^T$ where $\boldsymbol{q}_0^* \in \mathbb{R}^{n\times k_0}$ and $\boldsymbol{p}_0^* \in \mathbb{R}^{n\times n_0}$. Then, we have

$$\boldsymbol{q}_0^{*T}\boldsymbol{q}_0 = \mathbf{I}_{k_0},\ \boldsymbol{q}_0^{*T}\boldsymbol{p}_0 = 0,\ \boldsymbol{p}_0^{*T}\boldsymbol{q}_0 = 0,\ \boldsymbol{p}_0^{*T}\boldsymbol{p}_0 = \mathbf{I}_{n_0}. \tag{3.2.2}$$

If the matrix pencil is of index-1 we can use the above bases to remove the redundancy from either (3.1.9) or (3.1.8) otherwise we take the following steps:

Step 0, if $\mu > 1$:
If we left and right multiply (3.2.1) by $\boldsymbol{p}_0^{*T}$ and $\boldsymbol{p}_0$, respectively, we obtain

$$\mathbf{I}_{n_0} = \mathbf{Z}_{q_0} + \mathbf{Z}_{p_0}, \tag{3.2.3}$$

with $\mathbf{Z}_{p_0} := \boldsymbol{p}_0^{*T} \sum_{\substack{i=2\\ \mu>2}}^{\mu-1} \Pi_{i-1}\mathbf{Q}_i\boldsymbol{p}_0 + \boldsymbol{p}_0^{*T}\Pi_{\mu-1}\boldsymbol{p}_0$, $\mathbf{Z}_{q_0} := \boldsymbol{p}_0^{*T}\Pi_0\mathbf{Q}_1\boldsymbol{p}_0$. It is easy to check that $\mathbf{Z}_{p_0}$ and $\mathbf{Z}_{q_0}$ are mutually orthogonal projectors, acting in $\mathbb{R}^{n_0}$ provided admissible projectors are used. If, we let $z_{p_0} \in \operatorname{Im}\mathbf{Z}_{q_0}$ and $z_{p_0} \in \operatorname{Im}\mathbf{Z}_{p_0}$, then $\mathbf{Z}_{p_0}z_{p_0} = z_{p_0}$, $\mathbf{Z}_{p_0}z_{q_0} = 0$, $\mathbf{Z}_{q_0}z_{p_0} = 0$, $\mathbf{Z}_{q_0}z_{q_0} = z_{q_0}$. Let $k_1 = \dim(\operatorname{Im}\mathbf{Z}_{q_0})$, and $n_1 = n_0 - k_1$, and let us consider a basis matrix $(z_{p_0}, z_{q_0}) \in \mathbb{R}^{n_0}$ made of n_1 independent columns of projection matrix $\mathbf{Z}_{p_0}$ and k_1 independent columns of the complementary projection matrix $\mathbf{Z}_{q_0}$. We denote by $(z_{p_0}^{*T}, z_{q_0}^{*T})^T$ the inverse of (z_{p_0}, z_{q_0}), such that

$$z_{p_0}^{*T}z_{p_0} = \mathbf{I}_{n_1},\ z_{p_0}^{*T}z_{q_0} = 0,\ z_{q_0}^{*T}z_{p_0} = 0,\ z_{q_0}^{*T}z_{q_0} = \mathbf{I}_{k_1}. \tag{3.2.4}$$

We can see that if $\mu = 2$ then the bases of the projector products $\{\mathbf{Q}_0, \Pi_0\mathbf{Q}_1, \Pi_1\}$ in (3.2.1) are $\{\boldsymbol{q}_0, \boldsymbol{p}_0 z_{q_0}, \boldsymbol{p}_0 z_{p_0}\}$, respectively.

Step 1, if $\mu > 2$:
If we left and right multiply (3.2.3) by $z_{p_0}^{*T}$ and z_{p_0}, respectively, we obtain

$$\mathbf{I}_{n_1} = \mathbf{Z}_{q_1} + \mathbf{Z}_{p_1}, \tag{3.2.5}$$

with $\mathbf{Z}_{p_1} := z_{p_0}^{*T}\boldsymbol{p}_0^{*T} \sum_{\substack{i=3\\ \mu>3}}^{\mu-1} \Pi_{i-1}\mathbf{Q}_i\boldsymbol{p}_0 z_{p_0} + z_{p_0}^{*T}\boldsymbol{p}_0^{*T}\Pi_{\mu-1}\boldsymbol{p}_0 z_{p_0}$, $\mathbf{Z}_{q_1} := z_{p_0}^{*T}\boldsymbol{p}_0^{*T}\Pi_1\mathbf{Q}_2\boldsymbol{p}_0 z_{p_0}$. We can also see that the projectors are mutually orthogonal projectors, acting in $\mathbb{R}^{n_1}$. If, we let $z_{p_1} \in \operatorname{Im}\mathbf{Z}_{q_1}$ and $z_{p_1} \in \operatorname{Im}\mathbf{Z}_{p_1}$. Let $k_2 = \dim(\operatorname{Im}\mathbf{Z}_{q_1})$, and $n_2 = n_1 - k_2$, and let us consider a basis matrix $(z_{p_1}, z_{q_1}) \in \mathbb{R}^{n_1}$ made of n_2 independent columns of projection matrix $\mathbf{Z}_{p_0}$ and k_2 independent columns of the complementary projection matrix $\mathbf{Z}_{q_1}$. We denote by $(z_{p_1}^{*}, z_{q_1}^{*})^T$ the inverse of (z_{p_1}, z_{q_1}), such that

$$z_{p_1}^{*T}z_{p_1} = \mathbf{I}_{n_2},\ z_{p_1}^{*T}z_{q_1} = 0,\ z_{q_1}^{*T}z_{p_1} = 0,\ z_{q_1}^{*T}z_{q_1} = \mathbf{I}_{k_2},\ z_{p_1}z_{p_1}^{*T} + z_{q_1}z_{q_1}^{*T} = \mathbf{I}_{n_1}.$$

We can also see that if $\mu = 3$ then the bases of the projector products $\{\mathbf{Q}_0, \Pi_0\mathbf{Q}_1, \Pi_1\mathbf{Q}_2, \Pi_2\}$ are $\{\boldsymbol{q}_0, \boldsymbol{p}_0 z_{q_0}, \boldsymbol{p}_0 z_{p_0} z_{q_1}, \boldsymbol{p}_0 z_{p_0} z_{p_1}\}$ respectively.
Step j, if $\mu > j+1$:

It's interesting to see that this process is an iterative process and the jth iteration leads to an identity matrix given by:

$$\mathbf{I}_{n_j} = \mathbf{Z}_{q_j} + \mathbf{Z}_{p_j}, \quad j = 1, \ldots, \mu - 2, \quad \mu > 2, \tag{3.2.6}$$

with $\mathbf{Z}_{q_j} := z_{p_{j-1}}^{*T} \ldots z_{p_0}^{*T} \boldsymbol{p}_0^{*T} \Pi_j \mathbf{Q}_{j+1} \boldsymbol{p}_0 z_{p_0} \ldots z_{p_{j-1}}$ and $\mathbf{Z}_{p_j} := z_{p_{j-1}}^{*T} \ldots z_{p_0}^{*T} \boldsymbol{p}_0^{*T} \left(\sum_{\substack{i=j+2 \\ i<\mu-2}}^{\mu-1} \Pi_{i-1} \mathbf{Q}_i + \Pi_{\mu-1} \right) \boldsymbol{p}_0 z_{p_0} \ldots z_{p_{i-1}}$. These projectors are also mutually orthogonal projectors, acting in $\mathbb{R}^{n_j}$. If, we let $z_{p_j} \in \operatorname{Im} \mathbf{Z}_{q_j}$ and $z_{p_j} \in \operatorname{Im} \mathbf{Z}_{p_j}$, then $\mathbf{Z}_{p_j} z_{p_j} = z_{p_j}, \quad \mathbf{Z}_{p_j} z_{q_j} = 0, \quad \mathbf{Z}_{q_j} z_{p_j} = 0, \quad \mathbf{Z}_{q_j} z_{q_j} = z_{q_j}$. Let $k_{j+1} = \dim(\operatorname{Im} \mathbf{Z}_{q_j})$, and $n_{j+1} = n_j - k_{j+1}$, and let us consider a basis matrix $(z_{p_j}, z_{q_j}) \in \mathbb{R}^{n_j}$ made of n_{j+1} independent columns of projection matrix $\mathbf{Z}_{p_j}$ and k_{j+1} independent columns of the complementary projection matrix $\mathbf{Z}_{q_j}$. We denote by $(z_{p_j}^{*T}, z_{q_j}^{*T})^T$ the inverse of (z_{p_j}, z_{q_j}), such that

$$z_{p_j}^{*T} z_{p_j} = \mathbf{I}_{n_{j+1}}, \; z_{p_j}^{*T} z_{q_j} = 0, \; z_{q_j}^{*T} z_{p_j} = 0, \; z_{q_j}^{*T} z_{q_j} = \mathbf{I}_{k_{j+1}}, \; z_{p_j} z_{p_j}^{*T} + z_{q_j} z_{q_j}^{*T} = \mathbf{I}_{n_j}.$$

Then, we can represent $\mathbf{Z}_{p_j}, \mathbf{Z}_{q_j}$ as $\mathbf{Z}_{p_j} = z_{p_j} z_{p_j}^{*T}, \quad \mathbf{Z}_{q_j} = z_{q_j} z_{q_j}^{*T}$, and we have

$$\mathbf{Z}_{p_j} z_{p_j} = z_{p_j}, \quad \mathbf{Z}_{p_j} z_{q_j} = 0, \quad \mathbf{Z}_{q_j} z_{p_j} = 0, \quad \mathbf{Z}_{q_j} z_{q_j} = z_{q_j}. \tag{3.2.7}$$

Hence the bases of the projector products $\{\mathbf{Q}_0, \Pi_0 \mathbf{Q}_1, \ldots, \Pi_{i-1} \mathbf{Q}_i, \ldots, \Pi_{\mu-1}\}$ in (3.2.1) are $\{\boldsymbol{q}_0, \boldsymbol{p}_0 z_{q_0}, \ldots, \boldsymbol{p}_0 z_{p_0} \ldots z_{p_{i-2}} z_{q_{i-1}}, \ldots, \boldsymbol{p}_0 z_{p_0} \ldots z_{p_{\mu-2}}\}$, $i = 2, \ldots, \mu - 1$, respectively. Thus we can now expand $\boldsymbol{x}$ with respect to these bases, obtaining,

$$\boldsymbol{x} = \boldsymbol{p}_0 z_{p_0} \ldots z_{p_{\mu-2}} \xi_p + \boldsymbol{q}_0 \xi_{q,0} + \boldsymbol{p}_0 z_{q_0} \xi_{q,1} + \sum_{i=2}^{\mu-1} \boldsymbol{p}_0 z_{p_0} \ldots z_{p_{i-2}} z_{q_{i-1}} \xi_{q,i}, \tag{3.2.8}$$

where $\xi_p \in \mathbb{R}^{n_{\mu-1}}$, $\xi_{q,i} \in \mathbb{R}^{k_i}$, $\xi_{q,0} \in \mathbb{R}^{k_i}$, $i = 0, \ldots, \mu - 1$ and with inversion expressions

$$\begin{aligned} &\xi_p = z_{p_{\mu-2}}^{*T} \ldots z_{p_0}^{*T} \boldsymbol{p}_0^{*T} \boldsymbol{x}_P, \quad \xi_{q,0} = \boldsymbol{q}_0^{*T} \boldsymbol{x}_{Q,0}, \quad \xi_{q,1} = z_{q_0}^{*T} \boldsymbol{p}_0^{*T} \boldsymbol{x}_{Q,1}, \\ &\xi_{q,i} = z_{q_{i-1}}^{*T} z_{p_{i-2}}^{*T} \ldots z_{p_0}^{*T} \boldsymbol{p}_0^{*T} \boldsymbol{x}_{Q,i}, \quad i = 2, \ldots, \mu - 1. \end{aligned} \tag{3.2.9}$$

Substituting the variables in (3.2.33) and (3.2.9) into (3.1.8) leads to modified decoupled system given by

$$\xi_p' = \mathbf{A}_p \xi_p + \mathbf{B}_p \boldsymbol{u}, \tag{3.2.10a}$$

$$\xi_{q,\mu-1} = \mathbf{A}_{q,\mu-1} \xi_p + \mathbf{B}_{q,\mu-1} \boldsymbol{u}, \tag{3.2.10b}$$

$$\xi_{q,i} = \mathbf{A}_{q,i} \xi_p + \mathbf{B}_{q,i} \boldsymbol{u} + \sum_{j=i+1}^{\mu-1} \mathbf{A}_{q_{i,j}} \xi_{q,j}', \quad i = \mu - 2, \ldots 2, \tag{3.2.10c}$$

$$\xi_{q,1} = \mathbf{A}_{q,1}\xi_p + \mathbf{B}_{q,1}\boldsymbol{u} + \sum_{\substack{j=2\\ \mu>2}}^{\mu-1} \mathbf{A}_{q_{1,j}}\xi'_{q,j}, \tag{3.2.10d}$$

$$\xi_{q,0} = \mathbf{A}_{q,0}\xi_p + \mathbf{B}_{q,0}\boldsymbol{u} + \sum_{j=1}^{\mu-1} \mathbf{A}_{q_{0,j}}\xi'_{q,j}, \tag{3.2.10e}$$

$$\mathbf{y} = \mathbf{C}_p^T\xi_p + \sum_{i=0}^{\mu-1} \mathbf{C}_{q,i}^T\xi_{q,i}, \tag{3.2.10f}$$

where

$$\begin{aligned}
&\mathbf{A}_p := \boldsymbol{z}_{\boldsymbol{p}_{\mu-2}}^{*T}\cdots\boldsymbol{z}_{\boldsymbol{p}_0}^{*T}\boldsymbol{p}_0^{*T}\mathbf{A}_P\boldsymbol{p}_0\boldsymbol{z}_{\boldsymbol{p}_0}\cdots\boldsymbol{z}_{\boldsymbol{p}_{\mu-2}} \in \mathbb{R}^{n_p\times n_p}, \mathbf{B}_p := \boldsymbol{z}_{\boldsymbol{p}_{\mu-2}}^{*T}\cdots\boldsymbol{z}_{\boldsymbol{p}_0}^{*T}\boldsymbol{p}_0^{*T}\mathbf{B}_P \in \mathbb{R}^{n_p\times m},\\
&\mathbf{A}_{q,\mu-1} := \boldsymbol{z}_{\boldsymbol{q}_{\mu-2}}^{*T}\boldsymbol{z}_{\boldsymbol{p}_{\mu-3}}^{*T}\cdots\boldsymbol{z}_{\boldsymbol{p}_0}^{*T}\boldsymbol{p}_0^{*T}\mathbf{A}_{Q,\mu-1}\boldsymbol{p}_0\boldsymbol{z}_{\boldsymbol{p}_0}\cdots\boldsymbol{z}_{\boldsymbol{p}_{\mu-2}} \in \mathbb{R}^{k_{\mu-1}\times n_p},\\
&\mathbf{B}_{q,\mu-1} := \boldsymbol{z}_{\boldsymbol{q}_{\mu-2}}^{*T}\boldsymbol{z}_{\boldsymbol{p}_{\mu-3}}^{*T}\cdots\boldsymbol{z}_{\boldsymbol{p}_0}^{*T}\boldsymbol{p}_0^{*T}\mathbf{B}_{Q,\mu-1} \in \mathbb{R}^{k_{\mu-1}\times m},\\
&\mathbf{A}_{q,i} := \boldsymbol{z}_{\boldsymbol{q}_{i-1}}^{*T}\boldsymbol{z}_{\boldsymbol{p}_{i-2}}^{*T}\cdots\boldsymbol{z}_{\boldsymbol{p}_0}^{*T}\boldsymbol{p}_0^{*T}\mathbf{A}_{Q,i}\boldsymbol{p}_0\boldsymbol{z}_{\boldsymbol{p}_0}\cdots\boldsymbol{z}_{\boldsymbol{p}_{\mu-2}} \in \mathbb{R}^{k_i\times n_p},\\
&\mathbf{B}_{q,i} := \boldsymbol{z}_{\boldsymbol{q}_{i-1}}^{*T}\boldsymbol{z}_{\boldsymbol{p}_{i-2}}^{*T}\cdots\boldsymbol{z}_{\boldsymbol{p}_0}^{*T}\boldsymbol{p}_0^{*T}\mathbf{B}_{Q,i} \in \mathbb{R}^{k_i\times m},\\
&\mathbf{A}_{q_{i,j}} := \boldsymbol{z}_{\boldsymbol{q}_{i-1}}^{*T}\boldsymbol{z}_{\boldsymbol{p}_{i-2}}^{*T}\cdots\boldsymbol{z}_{\boldsymbol{q}_0}^{*T}\boldsymbol{p}_0^{*T}\mathbf{A}_{\mathbf{Q}_{i,j}}\boldsymbol{p}_0\boldsymbol{z}_{\boldsymbol{p}_0}\cdots\boldsymbol{z}_{\boldsymbol{p}_{j-2}}\boldsymbol{z}_{\boldsymbol{q}_{j-1}} \in \mathbb{R}^{k_i\times k_j}\\
&\mathbf{A}_{q,1} := \boldsymbol{z}_{\boldsymbol{q}_0}^{*T}\boldsymbol{p}_0^{*T}\mathbf{A}_{Q,1}\boldsymbol{p}_0\boldsymbol{z}_{\boldsymbol{p}_0}\cdots\boldsymbol{z}_{\boldsymbol{p}_{\mu-2}} \in \mathbb{R}^{k_1\times n_p},\quad \mathbf{B}_{q,1} := \boldsymbol{z}_{\boldsymbol{q}_0}^{*T}\boldsymbol{p}_0^{*T}\mathbf{B}_{Q,1} \in \mathbb{R}^{k_1\times m},\\
&\mathbf{A}_{q_{1,j}} := \boldsymbol{z}_{\boldsymbol{p}_0}^{*T}\boldsymbol{p}_0^{*T}\mathbf{A}_{\mathbf{Q}_{1,j}}\boldsymbol{p}_0\boldsymbol{z}_{\boldsymbol{p}_0}\cdots\boldsymbol{z}_{\boldsymbol{p}_{j-2}}\boldsymbol{z}_{\boldsymbol{q}_{j-1}} \in \mathbb{R}^{k_1\times k_j},\\
&\mathbf{A}_{q,0} := \boldsymbol{q}_0^{*T}\mathbf{A}_{Q,0}\boldsymbol{p}_0\boldsymbol{z}_{\boldsymbol{p}_0}\cdots\boldsymbol{z}_{\boldsymbol{p}_{\mu-2}} \in \mathbb{R}^{k_0\times n_p},\quad \mathbf{B}_{q,0} := \boldsymbol{q}_0^{*T}\mathbf{B}_{Q,0} \in \mathbb{R}^{k_0\times m},\\
&\mathbf{A}_{q_{0,j}} := \begin{cases} \boldsymbol{q}_0^{*T}\mathbf{A}_{\mathbf{Q}_{0,j}}\boldsymbol{p}_0\boldsymbol{z}_{\boldsymbol{q}_0}, & \text{If } j=1,\\ \boldsymbol{q}_0^{*T}\mathbf{A}_{\mathbf{Q}_{0,j}}\boldsymbol{p}_0\boldsymbol{z}_{\boldsymbol{p}_0}\cdots\boldsymbol{z}_{\boldsymbol{p}_{j-2}}\boldsymbol{z}_{\boldsymbol{q}_{j-1}}, & \text{Otherwise.}\end{cases}\\
&\mathbf{C}_p^T = \mathbf{C}^T\boldsymbol{p}_0\boldsymbol{z}_{\boldsymbol{p}_0}\cdots\boldsymbol{z}_{\boldsymbol{p}_{\mu-2}} \in \mathbb{R}^{\ell\times n_p},\ \mathbf{C}_{q,0}^T = \mathbf{C}^T\boldsymbol{q}_0 \in \mathbb{R}^{\ell\times k_0},\ \mathbf{C}_{q,1}^T = \mathbf{C}^T\boldsymbol{p}_0\boldsymbol{z}_{\boldsymbol{q}_0} \in \mathbb{R}^{\ell\times k_1},\\
&\mathbf{C}_{q,i}^T = \mathbf{C}^T\boldsymbol{p}_0\boldsymbol{z}_{\boldsymbol{p}_0}\cdots\boldsymbol{z}_{\boldsymbol{p}_{i-2}}\boldsymbol{z}_{\boldsymbol{q}_{i-1}} \in \mathbb{R}^{\ell\times k_i}\quad n_p = n_{\mu-1}.
\end{aligned}$$

We can observe that, (3.2.10) can be written in a compact form given by

$$\xi'_p = \mathbf{A}_p\xi_p + \mathbf{B}_p\mathbf{u}, \tag{3.2.11a}$$

$$-\mathcal{L}\xi'_q = \mathbf{A}_q\xi_p - \xi_q + \mathbf{B}_q\mathbf{u}, \tag{3.2.11b}$$

$$\mathbf{y} = \mathbf{C}_p^T\,\xi_p + \mathbf{C}_q^T\,\xi_q, \tag{3.2.11c}$$

where $\xi_p \in \mathbb{R}^{n_p}$, $\mathbf{A}_p \in \mathbb{R}^{n_p,n_p}$, $\mathbf{B}_p \in \mathbb{R}^{n_p,m}$. $\xi_q = (\xi_{q,\mu-1},\ldots,\xi_{q,0})^T \in \mathbb{R}^{n_q}$, $\mathbf{A}_q = (\mathbf{A}_{q,\mu-1},\ldots,\mathbf{A}_{q,0})^T \in \mathbb{R}^{n_q\times n_p}$, $\mathbf{B}_q = (\mathbf{B}_{q,\mu-1},\ldots,\mathbf{B}_{q,0})^T \in \mathbb{R}^{n_q\times m}$, $\mathbf{C}_q = (\mathbf{C}_{q,\mu-1}^T,\ldots,\mathbf{C}_{q,0}^T)^T \in \mathbb{R}^{n_q\times\ell}$ and $\mathcal{L} \in \mathbb{R}^{n_q\times n_q}$ is a strictly lower triangular nilpotent matrix of index-μ with entries $\mathbf{A}_{q_{i,j}}$ as defined in the decoupled system (3.2.10). n_p

and $n_q = \sum_{i=0}^{\mu-1} k_i$ are the number of differential and algebraic equations, respectively, and $n = n_p + n_q$ is the dimension of the DAE. Thus, the decoupled system (3.2.11) preserves the dimension of the DAE (2.3.1). It can be proved that the decoupled system (3.2.11) can be written in the form:

$$\xi_q = \sum_{i=0}^{\mu-1} \mathcal{L}^i \mathbf{A}_q \mathbf{A}_p^i \xi_p + \sum_{i=1}^{\mu-1} \sum_{k=0}^{i-1} \mathcal{L}^i \mathbf{A}_q \mathbf{A}_p^k \mathbf{B}_p \boldsymbol{u}^{(i-k-1)} + \sum_{i=0}^{\mu-1} \mathcal{L}^i \mathbf{B}_q \boldsymbol{u}^{(i)}, \tag{3.2.12}$$

where $\boldsymbol{u}^{(i)} \in \mathbb{R}^m$ is the ith derivative of the input data.

Next, we analyze the initial value of the DAE (2.3.1). Using system (3.2.11), we have: $\xi(0) := \begin{pmatrix} \xi_p(0) \\ \xi_q(0) \end{pmatrix}$, where

$$\xi_q(0) = \sum_{i=0}^{\mu-1} \mathcal{L}^i \mathbf{A}_q \mathbf{A}_p^i \xi_p(0) + \sum_{i=1}^{\mu-1} \sum_{k=0}^{i-1} \mathcal{L}^i \mathbf{A}_q \mathbf{A}_p^k \mathbf{B}_p \boldsymbol{u}^{(i-k-1)}(0) + \sum_{i=0}^{\mu-1} \mathcal{L}^i \mathbf{B}_q \boldsymbol{u}^{(i)}(0). \tag{3.2.13}$$

We observe that $\xi_p(0)$ can be chosen arbitrary while $\xi_q(0)$ has to be chosen such that the hidden constraint (3.2.13) is satisfied. Thus the initial value $\boldsymbol{x}(0)$ of DAE (2.3.1) has to be consistent and the input data has to be at least $\mu - 1$ times differentiable. In this approach we take care of this. In fact, if we apply initial condition $\xi_p(0) = z_{p_{\mu-2}}^{*T} \cdots z_{p_0}^{*T} p_0^{*T} \boldsymbol{x}(0)$, where $\boldsymbol{x}(0)$ is a consistent initial condition, we can solve the system (3.2.11) hierarchically by numerical integration of the differential part (3.2.11a) and then by computing the algebraic solutions using (3.2.12). Then the desired output solution is obtained by using (3.2.11c). It can be proved that $\sigma_f(\mathbf{E}, \mathbf{A}) = \sigma(\mathbf{A}_p)$, thus system (3.2.11) also preserves the stability of the DAE. The number of differential equation is always equal to the total algebraic multiplicity of the finite eigenvalues of the matrix pencil $(\mathbf{E}, \mathbf{A})$.

3.2.2 *Matrix Pencil* $(\mathbf{E}, \mathbf{A})$ *with Only Infinite Eigenvalue*

Here, we assume that the spectrum of the matrix pencil of (2.3.1) has no finite eigenvalues, this implies $\Pi_{\mu-1} = 0$. Thus, (3.2.1), reduces to

$$\mathbf{I}_n = \mathbf{Q}_0 + \sum_{i=1}^{\mu-1} \Pi_{i-1} \mathbf{Q}_i, \quad \mu > 1. \tag{3.2.14}$$

It is easy to see that if $\mu = 2$, then (3.2.14) simplifies to, $\mathbf{I}_n = \mathbf{Q}_0 + \Pi_0\mathbf{Q}_1 = \mathbf{Q}_0 + \mathbf{P}_0$. Thus, if $\mu = 2$ then the bases of the projector products $\{\mathbf{Q}_0, \Pi_0\mathbf{Q}_1\}$ in (3.2.14) are $\{\boldsymbol{q}_0, \boldsymbol{p}_0\}$, respectively otherwise we follow the steps below:

Step 0, if $\mu > 2$:
By construction, (3.2.14) can be written as:

$$\mathbf{I}_{n_0} = \boldsymbol{p}_0^{*T}\Pi_0\mathbf{Q}_1\boldsymbol{p}_0 + \boldsymbol{p}_0^{*T}\sum_{i=2}^{\mu-1}\Pi_{i-1}\mathbf{Q}_i\boldsymbol{p}_0.$$

Then,

$$\mathbf{I}_{n_0} = \mathbf{Z}_{q_0} + \mathbf{Z}_{p_0}, \tag{3.2.15}$$

with $\mathbf{Z}_{p_0} := \boldsymbol{p}_0^{*T}\sum_{\substack{i=2\\ \mu>2}}^{\mu-1}\Pi_{i-1}\mathbf{Q}_i\boldsymbol{p}_0$, $\mathbf{Z}_{q_0} := \boldsymbol{p}_0^{*T}\Pi_0\mathbf{Q}_1\boldsymbol{p}_0$. Here, also $\mathbf{Z}_{p_0}$ and $\mathbf{Z}_{q_0}$ are mutually orthogonal projectors, acting in $\mathbb{R}^{n_0}$. If we let $z_{p_0} \in \operatorname{Im}\mathbf{Z}_{q_0}$ and $z_{p_0} \in \operatorname{Im}\mathbf{Z}_{p_0}$, then $\mathbf{Z}_{p_0}z_{p_0} = z_{p_0}$, $\mathbf{Z}_{p_0}z_{q_0} = 0$, $\mathbf{Z}_{q_0}z_{p_0} = 0$, $\mathbf{Z}_{q_0}z_{q_0} = z_{q_0}$. Let $k_1 = \dim(\operatorname{Im}\mathbf{Z}_{q_0})$, and $n_1 = n_0 - k_1$, and let us consider a basis matrix $(z_{p_0}, z_{q_0}) \in \mathbb{R}^{n_0}$ made of n_1 independent columns of projection matrix $\mathbf{Z}_{p_0}$ and k_1 independent columns of the complementary projection matrix $\mathbf{Z}_{q_0}$. We denote by $(z_{p_0}^{*T}, z_{q_0}^{*T})^T$ the inverse of (z_{p_0}, z_{q_0}), such that

$$z_{p_0}^{*T}z_{p_0} = \mathbf{I}_{n_1},\ z_{p_0}^{*T}z_{q_0} = 0,\ z_{q_0}^{*T}z_{p_0} = 0,\ z_{q_0}^{*T}z_{q_0} = \mathbf{I}_{k_1}. \tag{3.2.16}$$

Thus, if $\mu = 3$, then the bases of the projector products $\{\mathbf{Q}_0, \Pi_0\mathbf{Q}_1, \Pi_1\mathbf{Q}_2\}$ in (3.2.14) are $\{\boldsymbol{q}_0, \boldsymbol{p}_0 z_{q_0}, \boldsymbol{p}_0 z_{p_0}\}$, respectively.

Step 1, if $\mu > 3$:
If we left and right multiply (3.2.15) by $z_{p_0}^{*T}$ and z_{p_0}, respectively, we obtain

$$\mathbf{I}_{n_1} = \mathbf{Z}_{q_1} + \mathbf{Z}_{p_1}, \tag{3.2.17}$$

with $\mathbf{Z}_{p_1} := z_{p_0}^{*T}\boldsymbol{p}_0^{*T}\sum_{\substack{i=3\\ \mu>3}}^{\mu-1}\Pi_{i-1}\mathbf{Q}_i\boldsymbol{p}_0 z_{p_0}$, $\mathbf{Z}_{q_1} := z_{p_0}^{*T}\boldsymbol{p}_0^{*T}\Pi_1\mathbf{Q}_2\boldsymbol{p}_0 z_{p_0}$. We can also see that the projectors are mutually orthogonal projectors, acting in $\mathbb{R}^{n_1}$.

If we let $z_{p_1} \in \operatorname{Im}\mathbf{Z}_{q_1}$ and $z_{p_1} \in \operatorname{Im}\mathbf{Z}_{p_1}$, then $\mathbf{Z}_{p_1}z_{p_1} = z_{p_1}$, $\mathbf{Z}_{p_1}z_{q_1} = 0$, $\mathbf{Z}_{q_1}z_{p_1} = 0$, $\mathbf{Z}_{q_1}z_{q_1} = z_{q_1}$. Let $k_2 = \dim(\operatorname{Im}\mathbf{Z}_{q_1})$, and $n_2 = n_1 - k_2$, and let us consider a basis matrix $(z_{p_1}, z_{q_1}) \in \mathbb{R}^{n_1}$ made of n_2 independent columns of projection matrix $\mathbf{Z}_{p_0}$ and k_2 independent columns of the complementary projection matrix $\mathbf{Z}_{q_1}$. We denote by $(z_{p_1}^{*T}, z_{q_1}^{*T})^T$ the inverse of (z_{p_1}, z_{q_1}), such that

$$z_{p_1}^{*T}z_{p_1} = \mathbf{I}_{n_2},\ z_{p_1}^{*T}z_{q_1} = 0,\ z_{q_1}^{*T}z_{p_1} = 0,\ z_{q_1}^{*T}z_{q_1} = \mathbf{I}_{k_2}.$$

Thus, if $\mu = 4$ then the bases of the projector products $\{\mathbf{Q}_0, \Pi_0\mathbf{Q}_1, \Pi_1\mathbf{Q}_2, \Pi_2\mathbf{Q}_3\}$ in (3.2.14) are $\{\boldsymbol{q}_0, \boldsymbol{p}_0 z_{q_0}, \boldsymbol{p}_0 z_{p_0} z_{q_1}, \boldsymbol{p}_0 z_{p_0} z_{p_1}\}$, respectively. This a recursive process which can easily be generalized.

Step j, if $\mu > j + 2$:

The jth iteration leads to an identity matrix given by:

$$\mathbf{I}_{n_j} = \mathbf{Z}_{q_j} + \mathbf{Z}_{p_j}, \quad j = 1, \ldots, \mu - 2, \quad \mu > 2, \tag{3.2.18}$$

with $\mathbf{Z}_{p_j} := z_{p_{j-1}}^{*T} \ldots z_{p_0}^{*T} \boldsymbol{p}_0^{*T} \sum_{\substack{i=j+2 \\ j<\mu-2}}^{\mu-1} \Pi_{i-1}\mathbf{Q}_i \boldsymbol{p}_0 z_{p_0} \ldots z_{p_{j-1}}$,

$\mathbf{Z}_{q_j} := z_{p_{j-1}}^{*T} \ldots z_{p_0}^{*T} \boldsymbol{p}_0^{*T} \Pi_j \mathbf{Q}_{j+1} \boldsymbol{p}_0 z_{p_0} \ldots z_{p_{j-1}}$. These projectors are also mutually orthogonal projectors, acting in $\mathbb{R}^{n_j}$. If we let $z_{p_j} \in \operatorname{Im} \mathbf{Z}_{q_j}$ and $z_{p_j} \in \operatorname{Im} \mathbf{Z}_{p_j}$, then $\mathbf{Z}_{p_j} z_{p_j} = z_{p_j}$, $\mathbf{Z}_{p_j} z_{q_j} = 0$, $\mathbf{Z}_{q_j} z_{p_j} = 0$, $\mathbf{Z}_{q_j} z_{q_j} = z_{q_j}$. Let $k_{j+1} = \dim(\operatorname{Im} \mathbf{Z}_{q_j})$, and $n_{j+1} = n_j - k_{j+1}$, and let us consider a basis matrix $(z_{p_j}, z_{q_j}) \in \mathbb{R}^{n_j}$ made of n_{j+1} independent columns of projection matrix $\mathbf{Z}_{p_j}$ and k_{j+1} independent columns of the complementary projection matrix $\mathbf{Z}_{q_j}$. We denote by $(z_{p_j}^{*T}, z_{q_j}^{*T})^T$ the inverse of (z_{p_j}, z_{q_j}), such that

$$z_{p_j}^{*T} z_{p_j} = \mathbf{I}_{n_{j+1}}, \quad z_{p_j}^{*T} z_{q_j} = 0, \quad z_{q_j}^{*T} z_{p_j} = 0, \quad z_{q_j}^{*T} z_{q_j} = \mathbf{I}_{k_{j+1}}.$$

Hence the bases of the projector products $\{\mathbf{Q}_0, \Pi_0\mathbf{Q}_1, \Pi_1\mathbf{Q}_2, \ldots, \Pi_{i-1}\mathbf{Q}_i, \ldots, \Pi_{\mu-2}\mathbf{Q}_{\mu-1}\}$ in (3.2.14) are $\{\boldsymbol{q}_0, \boldsymbol{p}_0 z_{q_0}, \boldsymbol{p}_0 z_{p_0} z_{q_1}, \ldots, \boldsymbol{p}_0 z_{p_0} \ldots z_{p_{i-3}} z_{q_{i-2}}, \ldots, \boldsymbol{p}_0 z_{p_0} \ldots z_{p_{\mu-2}}\}$, $i = 3, \ldots, \mu - 1$, respectively. Thus, we can now expand $\boldsymbol{x}$ with respect to these bases, obtaining,

$$\boldsymbol{x} = \boldsymbol{q}_0 \xi_{q,0} + \boldsymbol{p}_0 z_{q_0} \xi_{q,1} + \sum_{i=2}^{\mu-2} \boldsymbol{p}_0 z_{p_0} \ldots z_{p_{i-2}} z_{q_{i-1}} \xi_{q,i} + \boldsymbol{p}_0 z_{p_0} \ldots z_{p_{\mu-3}} \xi_{q,\mu-1}, \tag{3.2.19}$$

where $\xi_{\mu-1} \in \mathbb{R}^{n_{\mu-2}}$, $\xi_{q,i} \in \mathbb{R}^{k_i}$, $\xi_{q,0} \in \mathbb{R}^{k_i}$, $i = 0, \ldots, \mu - 2$ and with inversion expressions

$$\begin{aligned}
&\xi_{\mu-1} = z_{p_{\mu-3}}^{*T} \ldots z_{p_0}^{*T} \boldsymbol{p}_0^{*T} \boldsymbol{x}_{Q,\mu-1}, \quad \xi_{q,0} = \boldsymbol{q}_0^{*T} \boldsymbol{x}_{Q,0}, \quad \xi_{q,1} = z_{q_0}^{*T} \boldsymbol{p}_0^{*T} \boldsymbol{x}_{Q,1}, \\
&\xi_{q,i} = z_{q_{i-1}}^{*T} z_{p_{i-2}}^{*T} \ldots z_{p_0}^{*T} \boldsymbol{p}_0^{*T} \boldsymbol{x}_{Q,i}, \quad i = 2, \ldots, \mu - 2.
\end{aligned} \tag{3.2.20}$$

If we substitute the variables in (3.2.19) and (3.2.20) into (3.1.8) leads to modified decoupled system given by

$$\xi_{q,\mu-1} = \mathbf{B}_{q,\mu-1} \boldsymbol{u}, \tag{3.2.21a}$$

$$\xi_{q,i} = \mathbf{B}_{q,i} \boldsymbol{u} + \sum_{j=i+1}^{\mu-1} \mathbf{A}_{q_{i,j}} \xi'_{q,j}, \quad i = \mu - 2, \ldots 2, \tag{3.2.21b}$$

$$\xi_{q,1} = \mathbf{B}_{q,1}\boldsymbol{u} + \sum_{\substack{j=2\\ \mu>2}}^{\mu-1} \mathbf{A}_{q_{1,j}}\xi'_{q,j}, \tag{3.2.21c}$$

$$\xi_{q,0} = \mathbf{B}_{q,0}\boldsymbol{u} + \sum_{j=1}^{\mu-1} \mathbf{A}_{q_{0,j}}\xi'_{q,j}, \tag{3.2.21d}$$

$$\mathbf{y} = \sum_{i=0}^{\mu-1} \mathbf{C}^T_{q,i}\xi_{q,i}, \tag{3.2.21e}$$

where

$$\mathbf{B}_{q,\mu-1} := z^{*T}_{p_{\mu-3}} \cdots z^{*T}_{p_0} p^{*T}_0 \mathbf{B}_{Q,\mu-1} \in \mathbb{R}^{k_{\mu-1}\times m}, \quad \mathbf{B}_{q,i} := z^{*T}_{q_{i-1}} z^{*T}_{p_{i-2}} \cdots z^{*T}_{p_0} p^{*T}_0 \mathbf{B}_{Q,i} \in \mathbb{R}^{k_i\times m},$$

$$\mathbf{B}_{q,0} := q^{*T}_0 \mathbf{B}_{Q,0} \in \mathbb{R}^{k_0,m}, \quad \mathbf{B}_{q,1} := z^{*T}_{q_0} p^{*T}_0 \mathbf{B}_{Q,1} \in \mathbb{R}^{k_1\times m},$$

$$\mathbf{A}_{q_{i,j}} := \begin{cases} z^{*T}_{q_{i-1}} z^{*T}_{p_{i-2}} \cdots z^{*T}_{p_0} p^{*T}_0 \mathbf{A}_{\mathbf{Q}_{i,j}} p_0 z_{p_0} \cdots z_{p_{\mu-3}}, & \text{If } j = \mu-1, \\ z^{*T}_{q_{i-1}} z^{*T}_{p_{i-2}} \cdots z^{*T}_{p_0} p^{*T}_0 \mathbf{A}_{\mathbf{Q}_{i,j}} p_0 z_{p_0} \cdots z_{p_{j-2}} z_{q_{j-1}}, & \text{Otherwise,} \end{cases}$$

$$\mathbf{A}_{q_{1,j}} := \begin{cases} z^{*T}_{q_0} p^{*T}_0 \mathbf{A}_{\mathbf{Q}_{1,j}} p_0 z_{p_0} \cdots z_{p_{\mu-3}}, & \text{If, } j = \mu-1, \\ z^{*T}_{q_0} p^{*T}_0 \mathbf{A}_{\mathbf{Q}_{1,j}} p_0 z_{p_0} \cdots z_{p_{j-2}} z_{q_{j-1}}, & \text{Otherwise.} \end{cases}$$

$$\mathbf{A}_{q_{0,j}} := \begin{cases} q^{*T}_0 \mathbf{A}_{\mathbf{Q}_{0,j}} p_0 z_{q_0}, & \text{If, } j = 1, \\ q^{*T}_0 \mathbf{A}_{\mathbf{Q}_{0,j}} p_0 z_{p_0} \cdots z_{p_{j-2}} z_{q_{j-1}}, & \text{If, } 2 \le j \le \mu-2, \\ q^{*T}_0 \mathbf{A}_{\mathbf{Q}_{0,j}} p_0 z_{p_0} \cdots z_{p_{\mu-3}}, & \text{If, } j = \mu-1. \end{cases}$$

$$\mathbf{C}^T_{q,0} = \mathbf{C}^T q_0 \in \mathbb{R}^{\ell\times k_0},\ \mathbf{C}^T_{q,1} = \mathbf{C}^T p_0 z_{q_0} \in \mathbb{R}^{\ell\times k_1},\ \mathbf{C}^T_{q,i} = \mathbf{C}^T p_0 z_{p_0} \cdots z_{p_{i-2}} z_{q_{i-1}} \in \mathbb{R}^{\ell\times k_i}.$$

We can observe that Eq. (3.2.21) can be written as

$$-\mathcal{L}\xi'_q = -\xi_q + \mathbf{B}_q\mathbf{u} \tag{3.2.22a}$$

$$\mathbf{y} = \mathbf{C}^T_q\xi_q, \tag{3.2.22b}$$

where $\xi_q = (\xi_{q,\mu-1}, \ldots, \xi_{q,0})^T \in \mathbb{R}^n$, $\mathbf{B}_q = (\mathbf{B}_{q,\mu-1}, \ldots, \mathbf{B}_{q,0})^T \in \mathbb{R}^{n\times m}$, $\mathbf{C}_q = (\mathbf{C}^T_{q,\mu-1}, \ldots, \mathbf{C}^T_{q,0})^T \in \mathbb{R}^{n\times\ell}$, $\mathcal{L} \in \mathbb{R}^{n\times n}$ is a strictly lower triangular nilpotent matrix of index μ. It can also be proved that the decoupled system (3.2.22) can be written in the form:

$$\mathbf{y} = \mathbf{C}^T_q \sum_{i=0}^{\mu-1} \mathcal{L}^i \mathbf{B}_q \boldsymbol{u}^{(i)}, \tag{3.2.23}$$

where $\boldsymbol{u}^{(i)} \in \mathbb{R}^m$ is the ith derivative of the input data. We observe that, we have only algebraic equations and their solutions can be computed exactly. We can also observe that $n = \sum_{i=0}^{\mu-1} k_i$ is the total number of algebraic equations which is also equal to the dimension of the DAE. Thus the decoupled system (3.2.22) preserves the dimension of the DAE.

For comparison with the DAE (2.3.1), we can rewrite either system (3.2.11) or (3.2.22) in the descriptor form given by

$$\tilde{\mathbf{E}}\xi' = \tilde{\mathbf{A}}\xi + \tilde{\mathbf{B}}\mathbf{u}, \tag{3.2.24a}$$

$$\mathbf{y} = \tilde{\mathbf{C}}^T\xi, \tag{3.2.24b}$$

where: if the spectrum of the matrix pencil $(\mathbf{E}, \mathbf{A})$ has at least one finite eigenvalue, then $\tilde{\mathbf{E}} = \begin{pmatrix} \mathrm{I} & 0 \\ 0 & -\mathcal{L} \end{pmatrix} \in \mathbb{R}^{n\times n}$, $\tilde{\mathbf{A}} = \begin{pmatrix} \mathbf{A}_p & 0 \\ \mathbf{A}_q & -\mathrm{I} \end{pmatrix} \in \mathbb{R}^{n\times n}$, $\tilde{\mathbf{B}} = \begin{pmatrix} \mathbf{B}_p \\ \mathbf{B}_q \end{pmatrix} \in \mathbb{R}^{n\times m}$, $\tilde{\mathbf{C}} = \begin{pmatrix} \mathbf{C}_p \\ \mathbf{C}_q \end{pmatrix} \in \mathbb{R}^{n\times \ell}$ and if the spectrum of the matrix pencil $(\mathbf{E}, \mathbf{A})$ has no finite eigenvalue, then $\tilde{\mathbf{E}} = -\mathcal{L} \in \mathbb{R}^{n\times n}$, $\tilde{\mathbf{A}} = -\mathrm{I} \in \mathbb{R}^{n\times n}$, $\tilde{\mathbf{B}} = \mathbf{B}_q \in \mathbb{R}^{n\times m}$, $\tilde{\mathbf{C}} = \mathbf{C}_q \in \mathbb{R}^{n\times \ell}$. We note that if we use canonical projectors in advance $\mathbf{A}_q = 0$ always, that it is, $\tilde{\mathbf{A}} = \begin{pmatrix} \mathbf{A}_p & 0 \\ 0 & -\mathrm{I} \end{pmatrix} \in \mathbb{R}^{n\times n}$. We can observe that this form reveals the interconnection structure of the DAE (2.3.1). Moreover it can be proved that systems (2.3.1) and (3.2.24) are equivalent. This implies that also their respective matrix pencils $(\mathbf{E}, \mathbf{A})$ and $(\tilde{\mathbf{E}}, \tilde{\mathbf{A}})$ are equivalent. If we consider DAEs whose matrix pencil $(\mathbf{E}, \mathbf{A})$ has at least one finite eigenvalue, we can show that they have same spectrum, since we can easily show that $\det(\lambda\tilde{\mathbf{E}} - \tilde{\mathbf{A}}) = \det(\lambda\mathrm{I} - \mathbf{A}_p)$, since $\det(\mathrm{I} - \lambda\mathcal{L}) = (1)^{n_q}$. This identity shows that the finite eigenvalues of the matrix pencil $(\mathbf{E}, \mathbf{A})$ coincide with the (possibly complex) eigenvalues of the matrix $\mathbf{A}_p$ of the differential part, which are exactly n_p, counting their multiplicity, i.e., $\sigma(\mathbf{A}_p) = \sigma_f(\mathbf{E}, \mathbf{A})$. Thus, the differential part of the decoupled system inherits the stability properties of DAEs.

Examples 3.2.1–3.2.5, illustrate how one can decouple index-1 to index-3 DAEs using matrix, projector and basis chains. We also illustrate the effect of using canonical projectors. Examples 3.2.1–3.2.3 originates from [6] while the rest originates from [4].

Example 3.2.1 Consider a linear RLC electric network in Fig. 3.1. We need to find the unknowns $x = \begin{pmatrix} e_1 & e_2 & e_3 & \imath_L & \imath_V \end{pmatrix}^T$ in this electric network. Modeling of this electrical network leads to a DAE of the form (2.3.1) with system matrices:

$$\mathbf{E} = \begin{pmatrix} 0&0&0&0&0 \\ 0&0&0&0&0 \\ 0&0&C&0&0 \\ 0&0&0&L&0 \\ 0&0&0&0&0 \end{pmatrix}, \quad \mathbf{A} = \begin{pmatrix} -G & G & 0 & 0 & 1 \\ G & -G & 0 & -1 & 0 \\ 0 & 0 & 0 & 1 & 0 \\ 0 & 1 & -1 & 0 & 0 \\ -1 & 0 & 0 & 0 & 0 \end{pmatrix}, \quad \mathbf{B} = \begin{pmatrix} 0 \\ 0 \\ 0 \\ 0 \\ -1 \end{pmatrix}, \quad \mathbf{C} = \mathbf{B}. \tag{3.2.25}$$

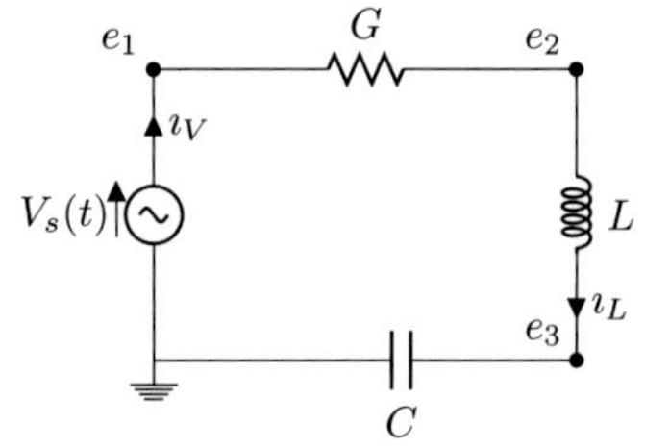

Fig. 3.1 Simple RLC electric network

This DAE has a regular matrix pencil since $\det(\lambda E - A) = GCL\lambda^2 + C\lambda + G \neq 0$ for some $\lambda \in \mathbb{C}$ and has two finite eigenvalues. Thus, we expect its decoupled system to have 2 differential and 3 algebraic equations. In order to decoupled this DAE, we need to first compute the matrix and projector chain using Definition 3.1.1. Setting $\mathbf{E}_0 = \mathbf{E}, \quad \mathbf{A}_0 = \mathbf{A}$, we can choose projector $\mathbf{Q}_0$ such that $\operatorname{Im}\mathbf{Q}_0 = \operatorname{Ker}\mathbf{E}_0$, and then compute $\mathbf{E}_1 = \mathbf{E}_0 - \mathbf{A}_0\mathbf{Q}_0$. Thus we obtain the first iterate of matrix and projector chain given by:

$$\mathbf{Q}_0 = \begin{pmatrix} 1&0&0&0&0\\ 0&1&0&0&0\\ 0&0&0&0&0\\ 0&0&0&0&0\\ 0&0&0&0&1 \end{pmatrix}, \quad \mathbf{P}_0 = \begin{pmatrix} 0&0&0&0&0\\ 0&0&0&0&0\\ 0&0&1&0&0\\ 0&0&0&1&0\\ 0&0&0&0&0 \end{pmatrix}, \quad \mathbf{E}_1 = \begin{pmatrix} G&-G&0&0&-1\\ -G&G&0&0&0\\ 0&0&C&0&0\\ 0&-1&0&L&0\\ 1&0&0&0&0 \end{pmatrix}. \tag{3.2.26}$$

Since $\mathbf{E}_1$ is nonsingular, thus Eq. (3.2.25) is an index-1 system. Next, we need to construct the basis chain, this is done following procedure in Sect. 3.2.1. Thus, we have

$$\boldsymbol{q}_0 = \begin{pmatrix} 1&0&0\\ 0&1&0\\ 0&0&0\\ 0&0&0\\ 0&0&1 \end{pmatrix} \in \operatorname{Ker}\mathbf{E}_0, \quad \boldsymbol{p}_0 = \begin{pmatrix} 0&0\\ 0&0\\ 1&0\\ 0&1\\ 0&0 \end{pmatrix} \in \operatorname{Im}\mathbf{P}_0. \tag{3.2.27}$$

We can easily check the above bases indeed satisfy: $\mathbf{Q}_0\boldsymbol{q}_0 = \boldsymbol{q}_0$, $\mathbf{P}_0\boldsymbol{q}_0 = \boldsymbol{q}_0$, $\mathbf{P}_0\boldsymbol{p}_0 = \boldsymbol{p}_0$, $\mathbf{Q}_0\boldsymbol{p}_0 = \boldsymbol{q}_0$. Using these bases we can build an invertible matrix $\begin{pmatrix}\boldsymbol{p}_0 & \boldsymbol{q}_0\end{pmatrix}$.

Then, using the inverse of this basis matrix we can construct the the inverse of $\boldsymbol{p}_0$ and $\boldsymbol{q}_0$ as, $(\boldsymbol{p}_0\ \boldsymbol{q}_0)^{-1} = (\boldsymbol{p}_0^{*T}, \boldsymbol{q}_0^{*T})^T$. Partitioning this inverse matrix leads to

$$\mathbf{q}_0^* = \begin{pmatrix} 1\,0\,0 \\ 0\,1\,0 \\ 0\,0\,0 \\ 0\,0\,0 \\ 0\,0\,1 \end{pmatrix}, \quad \mathbf{p}_0^* = \begin{pmatrix} 0\,0 \\ 0\,0 \\ 1\,0 \\ 0\,1 \\ 0\,0 \end{pmatrix}.$$

Since this an index-1 system, thus substituting $\mu = 1$ into (3.2.10) and above matrix, projector and basis chain leads to leads to the decoupled system of system (3.2.25) given by

$$\xi_p' = \mathbf{A}_p\xi_p + \mathbf{B}_p\mathbf{u}, \quad \xi_p(0) = \begin{pmatrix} e_3(0) \\ \imath_L(0) \end{pmatrix}, \tag{3.2.28a}$$

$$\xi_{q,0} = \mathbf{A}_{q,0}\xi_p + \mathbf{B}_{q,0}\mathbf{u}, \tag{3.2.28b}$$

$$\mathbf{y} = \mathbf{C}_p\xi_p + \mathbf{C}_{q,0}\xi_{q,0}. \tag{3.2.28c}$$

with system matrices given by $\mathbf{A}_p = \begin{pmatrix} 0 & \frac{1}{C} \\ -\frac{1}{L} & -\frac{1}{GL} \end{pmatrix}$, $\mathbf{B}_p = \begin{pmatrix} 0 \\ -\frac{1}{L} \end{pmatrix}$, $\mathbf{A}_{q,0} = \begin{pmatrix} 0 & 0 \\ 0 & -\frac{1}{G} \\ 0 & 1 \end{pmatrix}$, $\mathbf{B}_{q,0} = \begin{pmatrix} -1 \\ -1 \\ 0 \end{pmatrix}$, $\mathbf{C}_p = \begin{pmatrix} 0 \\ 0 \end{pmatrix}$ and $\mathbf{C}_{q,0} = \begin{pmatrix} 0 & 0 & -1 \end{pmatrix}^T$. Observe that $n_0 = 2, k_0 = 3 \Rightarrow n = n_0 + k_0 = 5$ and $\sigma_f(\mathbf{E}, \mathbf{A}) = \sigma(\mathbf{A}_p)$ as expected. Thus system (3.2.25) is decoupled into 2 and 3 differential and algebraic equations, respectively. We can observe that the decoupled system (3.2.28) can be solved in hierarchical way if wc apply $\xi_p(0) = \left(e_3(0)^T\ \imath_L(0)^T\right)^T$.

From system (3.2.28), we can observe that there is still a coupling between the differential part (3.2.32a) and algebraic part (3.2.28b) since $\mathbf{A}_{q,0} \neq 0$. We have earlier discussed that, we can enforce a complete decoupling, that is $\mathbf{A}_{q,0}$ if we choose canonical projectors in advance. For comparison, we can now repeat the same steps (3.2.26)–(3.2.28) but this time using canonical projector. Using formula (3.1.10), we can construct a construct canonical projectors using matrices (3.2.26) as follows: Setting $\mathbf{Q}_0^{(0)=\mathbf{Q}_0}$, $\mathbf{E}_1^{(0)} = \mathbf{E}_1$, $\mathbf{A}_0^{(0)} = \mathbf{A}_0$, then, we have

$$\mathbf{Q}_0^{(1)} = -\mathbf{Q}_0^{(0)}\mathbf{E}_1^{(0)^{-1}}\mathbf{A}_0^{(0)} = \begin{pmatrix} 1\,0\,0 & 0 & 0 \\ 0\,1\,0 & 1/G & 0 \\ 0\,0\,0 & 0 & 0 \\ 0\,0\,0 & 0 & 0 \\ 0\,0\,0 & -1 & 1 \end{pmatrix} \quad \text{and} \quad \mathbf{P}_0^{(1)} = \begin{pmatrix} 0\,0\,0 & 0 & 0 \\ 0\,0\,0 & -1/G & 0 \\ 0\,0\,1 & 0 & 0 \\ 0\,0\,0 & 1 & 0 \\ 0\,0\,0 & 1 & 0 \end{pmatrix}.$$

It is easy to check that $\mathbf{Q}_0^{(1)}$ is indeed a canonical projector. For convenience, We can set $\mathbf{Q}_0 = \mathbf{Q}_0^{(1)}$ and we repeat steps (3.2.26)–(3.2.28) as follows. Then

$$\mathbf{E}_1 = \mathbf{E}_0 - \mathbf{A}_0\mathbf{Q}_0 = \begin{pmatrix} G & -G & 0 & 0 & -1 \\ -G & G & 0 & 1 & 0 \\ 0 & 0 & C & 0 & 0 \\ 0 & -1 & 0 & L-1/G & 0 \\ 1 & 0 & 0 & 0 & 0 \end{pmatrix} \tag{3.2.29}$$

The basis chain as given by

$$\boldsymbol{p}_0 = \begin{pmatrix} 0 & 0 \\ 0 & -1/G \\ 1 & 0 \\ 0 & 1 \\ 0 & 1 \end{pmatrix} \in \operatorname{Im}\mathbf{P}_0 \quad \text{and} \quad \boldsymbol{q}_0 = \begin{pmatrix} 1 & 0 & 0 \\ 0 & 1 & 0 \\ 0 & 0 & 0 \\ 0 & 0 & 0 \\ 0 & 0 & 1 \end{pmatrix} \in \operatorname{Ker}\mathbf{E}_0 \tag{3.2.30}$$

and their respective inverses given by

$$\boldsymbol{p}^* = \begin{pmatrix} 0 & 0 \\ 0 & 0 \\ 1 & 0 \\ 0 & 1 \\ 0 & 0 \end{pmatrix} \quad \text{and} \quad \boldsymbol{q}^* = \begin{pmatrix} 1 & 0 & 0 & 0 & 0 \\ 0 & 1 & 0 & 1/G & 0 \\ 0 & 0 & 0 & -1 & 1 \end{pmatrix}^T. \tag{3.2.31}$$

Substituting $\mu = 1$ into (3.2.10) and above matrix, projector and basis chain leads to leads to a completely decoupled system of system (3.2.25) given by

$$\xi_p' = \begin{pmatrix} 0 & \frac{1}{C} \\ -\frac{1}{L} & -\frac{1}{GL} \end{pmatrix} \xi_p + \begin{pmatrix} 0 \\ -\frac{1}{L} \end{pmatrix} \mathbf{u}, \quad \xi_p(0) = \begin{pmatrix} e_3(0) \\ \imath_L(0) \end{pmatrix}, \tag{3.2.32a}$$

$$\xi_{q,0} = \begin{pmatrix} 0 & 0 \\ 0 & 0 \\ 0 & 0 \end{pmatrix} \xi_p + \begin{pmatrix} -1 \\ -1 \\ 0 \end{pmatrix} \mathbf{u}, \tag{3.2.32b}$$

$$\mathbf{y} = \begin{pmatrix} 0 & -1 \end{pmatrix} \xi_p + \begin{pmatrix} 0 & 0 & -1 \end{pmatrix} \xi_{q,0}. \tag{3.2.32c}$$

If we compare systems (3.2.28) and (3.2.32), we can observe this time $\mathbf{A}_{q,0} = 0$ which allows a complete decoupling and $\mathbf{C}_p = \begin{pmatrix} 0 & -1 \end{pmatrix}^T$ is nonzero. However, both lead to the same output solution given by $\mathbf{y} = -\xi_{p_2}$.

Example 3.2.2 In this example, we consider another linear RLC electric network with system matrices:

$$\mathbf{E} = \begin{pmatrix} C & 0 & 0 & 0 \\ 0 & 0 & 0 & 0 \\ 0 & 0 & L & 0 \\ 0 & 0 & 0 & 0 \end{pmatrix}, \quad \mathbf{A} = \begin{pmatrix} -G & G & 0 & 1 \\ G & -G & -1 & 0 \\ 0 & 1 & 0 & 0 \\ -1 & 0 & 0 & 0 \end{pmatrix}, \quad \mathbf{B} = \begin{pmatrix} 0 \\ 0 \\ 0 \\ -1 \end{pmatrix}, \quad x = \begin{pmatrix} e_1 \\ e_2 \\ \imath_L \\ \imath_V \end{pmatrix}. \tag{3.2.33}$$

This is also DAE with system dimension $n = 4$ and it has a regular matrix pencil since $\det(\lambda E - A) = GL\lambda + 1 \neq 0$ for some $\lambda \in \mathbb{C}$. We construct matrix and projector chains by first setting $\mathbf{E}_0 = \mathbf{E}$, $\mathbf{A}_0 = \mathbf{A}$. It also has a finite spectrum since $\sigma_f(\mathbf{E}, \mathbf{A}) = \{-1/GL\} = \emptyset$. Thus it decoupled system will take the form (3.2.10). Following the procedure discussed in Sect. 3.2.1, we can decouple system (3.2.33) into differential and algebraic parts as follows. First we need to construct the matrix and projector chain using Definition 3.1.1. This is done as follows, We can choose projector $\mathbf{Q}_0$ such that $\operatorname{Im} \mathbf{Q}_0 = \operatorname{Ker} \mathbf{E}_0$ and its complementary projector $\mathbf{P}_0$ given by

$$\mathbf{Q}_0 = \begin{pmatrix} 0 & 0 & 0 & 0 \\ 0 & 1 & 0 & 0 \\ 0 & 0 & 0 & 0 \\ 0 & 0 & 0 & 1 \end{pmatrix}, \quad \text{and} \quad \mathbf{P}_0 = \begin{pmatrix} 1 & 0 & 0 & 0 \\ 0 & 0 & 0 & 0 \\ 0 & 0 & 1 & 0 \\ 0 & 0 & 0 & 0 \end{pmatrix}. \tag{3.2.34}$$

Then,

$$\mathbf{E}_1 = \mathbf{E}_0 - \mathbf{A}_0\mathbf{Q}_0 = \begin{pmatrix} C & -G & 0 & -1 \\ 0 & G & 0 & 0 \\ 0 & -1 & L & 0 \\ 0 & 0 & 0 & 0 \end{pmatrix}, \quad \mathbf{A}_1 = \mathbf{A}_0\mathbf{P}_0 = \begin{pmatrix} 0 & G & 0 & 1 \\ 0 & -G & 0 & 0 \\ 0 & 1 & 0 & 0 \\ 0 & 0 & 0 & 0 \end{pmatrix}.$$

Since $\mathbf{E}_1$ is singular, the index-1 condition is violeted. Thus we need to continue with the process and choose projector $\mathbf{Q}_1$ such that $\operatorname{Im} \mathbf{Q}_1 = \operatorname{Ker} \mathbf{E}_1$ and its complementary projector $\mathbf{P}_1 = \mathbf{I} - \mathbf{Q}_1$ given by

$$\mathbf{Q}_1 = \frac{1}{C^2+1} \begin{pmatrix} 1 & 0 & 0 & C \\ 0 & 0 & 0 & 0 \\ 0 & 0 & 0 & 0 \\ C & 0 & 0 & C^2 \end{pmatrix}, \quad \text{and} \quad \mathbf{P}_1 = \mathbf{I} - \mathbf{Q}_1. \tag{3.2.35}$$

Then,

$$\mathbf{E}_2 = \mathbf{E}_1 - \mathbf{A}_1\mathbf{Q}_1 = \begin{pmatrix} C + \frac{G}{C^2+1} & -G & 0 & -1 + \frac{GC}{C^2+1} \\ -\frac{G}{C^2+1} & G & 0 & -\frac{GC}{C^2+1} \\ 0 & -1 & L & 0 \\ \frac{1}{C^2+1} & 0 & 0 & -\frac{C}{C^2+1} \end{pmatrix}. \tag{3.2.36}$$

We can easily check that $\mathbf{E}_2$ is non-singular. Thus Eq. (3.2.33) is of index-2. As we discussed earlier that above matrix and projector chain cannot be used to decouple higher index DAEs since they are not admissible, that is $\mathbf{Q}_1\mathbf{Q}_0 \neq 0$. We can easily check notice that projectors (3.2.34) and (3.2.35) don't satisfy this condition. Next, we need to construct admissible projectors which satisfy condition $\mathbf{Q}_1\mathbf{Q}_0 = 0$ as follows. We can make the following update on the projector $\mathbf{Q}_1$.

$$\mathbf{Q}_1^{(1)} := -\mathbf{Q}_1\mathbf{E}_2^{-1}\mathbf{A}_1 = \begin{pmatrix} 1 & 0 & 0 & 0 \\ 0 & 0 & 0 & 0 \\ 0 & 0 & 0 & 0 \\ C & 0 & 0 & 0 \end{pmatrix}, \quad \text{and} \quad \mathbf{P}_1^{(1)} = \mathbf{I} - \mathbf{Q}_1^{(1)}.$$

We can easily check that condition $\mathbf{Q}_1^{(1)}\mathbf{Q}_0 = 0$ is now satisfied. Hence $\mathbf{Q}_1^{(1)}$ and $\mathbf{Q}_0$ are admissible projectors. We then use $\mathbf{Q}_1^{(1)}$ instead of $\mathbf{Q}_1$ in the decoupling process. If we repeat step (3.2.36), we obtain:

$$\mathbf{E}_2 = \mathbf{E}_1 - \mathbf{A}_1\mathbf{Q}_1^{(1)} = \begin{pmatrix} C+G & -G & 0 & -1 \\ -G & G & 0 & 0 \\ 0 & -1 & L & 0 \\ 1 & 0 & 0 & 0 \end{pmatrix}, \quad \mathbf{A}_2 = \mathbf{A}_1\mathbf{P}_1^{(1)} = \begin{pmatrix} 0 & 0 & 0 & 0 \\ 0 & 0 & -1 & 0 \\ 0 & 0 & 0 & 0 \\ 0 & 0 & 0 & 0 \end{pmatrix}. \tag{3.2.37}$$

For convenience we can set $\mathbf{Q}_1 = \mathbf{Q}_1^{(1)}$ and $\mathbf{P}_1 = \tilde{\mathbf{P}}_1$. Since Eq. (3.2.33) is an index-2 system and its matrix pencil has at least one finite eigenvalues. Thus its decouple system will be in form (3.2.10). Next step is to construct basis chain following procedure in Sect. 3.2.1. The basis matrices for projectors $\mathbf{P}_0$ and $\mathbf{Q}_0$ are given by

$$\mathbf{p}_0 = \begin{pmatrix} 1 & 0 \\ 0 & 0 \\ 0 & 1 \\ 0 & 0 \end{pmatrix} \in \operatorname{Im}\mathbf{P}_0 \quad \text{and} \quad \mathbf{q}_0 = \begin{pmatrix} 0 & 0 \\ 1 & 0 \\ 0 & 0 \\ 0 & 1 \end{pmatrix} \in \operatorname{Ker}\mathbf{E}_0 \tag{3.2.38}$$

and their respective inverses are given by

$$\mathbf{p}_0^* = \begin{pmatrix} 1 & 0 \\ 0 & 0 \\ 0 & 1 \\ 0 & 0 \end{pmatrix}, \quad \mathbf{q}_0^* = \begin{pmatrix} 0 & 0 \\ 1 & 0 \\ 0 & 0 \\ 0 & 1 \end{pmatrix}. \tag{3.2.39}$$

From above bases, we compute other set of projectors

$$\mathbf{Z}_{p_0} = \mathbf{p}_0^{*T}\mathbf{P}_1\mathbf{p}_0 = \begin{pmatrix} 0 & 0 \\ 0 & 1 \end{pmatrix}, \quad \text{and} \quad \mathbf{Z}_{q_0} = \mathbf{p}_0^{*T}\mathbf{Q}_1\mathbf{p}_0 = \begin{pmatrix} 1 & 0 \\ 0 & 0 \end{pmatrix}.$$

Next, we construct the basis matrices of these projectors and there respective inverses given by

$$\mathbf{z}_{p_0} = \begin{pmatrix} 0 \\ 1 \end{pmatrix} \in \operatorname{Im}\mathbf{Z}_{p_0}, \ \mathbf{z}_{q_0} = \begin{pmatrix} 1 \\ 0 \end{pmatrix} \in \operatorname{Im}\mathbf{Z}_{q_0}, \quad \text{and} \quad \mathbf{z}_{p_0}^* = \begin{pmatrix} 0 \\ 1 \end{pmatrix}, \quad \mathbf{z}_{q_0}^* = \begin{pmatrix} 1 \\ 0 \end{pmatrix}. \tag{3.2.40}$$

Setting $\mu = 2$ and substituting Eqs. (3.2.37)–(3.2.40) into (3.2.10) leads to the decoupled system of (3.2.33) given by

$$\begin{aligned} \xi_p' &= \mathbf{A}_p\xi_p + \mathbf{B}_p\mathbf{u}, \quad \xi_p(0) = \iota_L(0),\\ \xi_{q,1} &= \mathbf{A}_{q,1}\xi_p + \mathbf{B}_{q,1}\mathbf{u},\\ \xi_{q,0} &= \mathbf{A}_{q,0}\xi_p + \mathbf{B}_{q,0}\mathbf{u} + \mathbf{A}_{q_{0,1}}\xi_{q,1}',\\ \mathbf{y} &= \mathbf{C}_p^T\xi_p + \mathbf{C}_{q,1}\xi_{q,1} + \mathbf{C}_{q,0}\xi_{q,0}, \end{aligned} \tag{3.2.41}$$

with system matrices $\mathbf{A}_p = -\frac{1}{GL}$, $\mathbf{B}_p = -\frac{1}{L}$, $\mathbf{A}_{q,1} = 0$, $\mathbf{B}_{q,1} = -1$, $\mathbf{A}_{q,0} = \begin{pmatrix} -\frac{1}{G} \\ 1 \end{pmatrix}$, $\mathbf{B}_{q,0} = \begin{pmatrix} -1 \\ 0 \end{pmatrix}$, $\mathbf{A}_{q_{0,1}} = \begin{pmatrix} 0 \\ C \end{pmatrix}$, $\mathbf{C}_p = 0$, $\mathbf{C}_{q,1} = 0$ and $\mathbf{C}_{q,0} = \begin{pmatrix} 0 \\ -1 \end{pmatrix}$. We observe that $n_{10} = 1, k_1 = 1, k_0 = 2 \Rightarrow n = n_{10} + k_1 + k_0 = 4$ and $\sigma_f(\mathbf{E}, \mathbf{A}) = \sigma(\mathbf{A}_p) = \{-\frac{1}{GL}\}$. Equation (3.2.41) can be solved if we apply $\xi_p(0) = \iota_L(0)$. Thus the DAE system (3.2.33) is decoupled into 1 differential and 3 algebraic equations.

We can observe that the above decoupled system still has the a one way coupling between the differential and algebraic part since $\mathbf{A}_q = \begin{pmatrix} \mathbf{A}_{q,1} \\ \mathbf{A}_{q,0} \end{pmatrix}$ is nonzero. As, we discussed earlier in order ro get a completely decoupled system, we need to use canonical projectors in advance. For comparison, we decouple DAE (3.2.33) this time using canonical projectors as follows. Using formula (3.1.10), we can update the already constructed admissible projectors leading to canonical projectors given by

$$\mathbf{Q}_0^{(2)} = \begin{pmatrix} 0 & 0 & 0 & 0 \\ 0 & 1 & 1/G & 0 \\ 0 & 0 & 0 & 0 \\ 0 & 0 & -1 & 1 \end{pmatrix}, \quad \mathbf{Q}_1^{(2)} = \begin{pmatrix} 1 & 0 & 0 & 0 \\ 0 & 0 & 0 & 0 \\ 0 & 0 & 0 & 0 \\ C & 0 & 0 & 0 \end{pmatrix}$$

and their respective complementary projector $\mathbf{P}_i^{(2)} = \mathrm{I} - \mathbf{Q}_i^{(2)}$. For convenience, we can see $\mathbf{Q}_0 = \mathbf{Q}_0^{(2)}$ and $\mathbf{Q}_1 = \mathbf{Q}_1^{(2)}$, and repeat the same steps as before. Then,

$$\mathbf{E}_2 = \mathbf{E}_1 - \mathbf{A}_1\mathbf{Q}_1 = \begin{pmatrix} C+G & -G & 0 & -1 \\ -G & G & 1 & 0 \\ 0 & -1 & \frac{GL-1}{G} & 0 \\ 1 & 0 & 0 & 0 \end{pmatrix}, \quad \mathbf{A}_2 = \mathbf{A}_0\mathbf{P}_0\mathbf{P}_1 = \begin{pmatrix} 0 & 0 & 0 & 0 \\ 0 & 0 & 0 & 0 \\ 0 & 0 & -1/G & 0 \\ 0 & 0 & 0 & 0 \end{pmatrix} \tag{3.2.42}$$

Next, we construct the basis chain and their respective inverse as before, which are given by

$$\mathbf{p}_0 = \begin{pmatrix} 1 & 0 \\ 0 & -1/G \\ 0 & 1 \\ 0 & 1 \end{pmatrix}, \quad \mathbf{q}_0 = \begin{pmatrix} 0 & 0 \\ 1 & 0 \\ 0 & 0 \\ 0 & 1 \end{pmatrix}, \quad \text{and} \quad \mathbf{p}_0^* = \begin{pmatrix} 1 & 0 \\ 0 & 0 \\ 0 & 1 \\ 0 & 0 \end{pmatrix}, \quad \mathbf{q}_0^* = \begin{pmatrix} 0 & 0 \\ 1 & 0 \\ 1/G & -1 \\ 0 & 1 \end{pmatrix}. \tag{3.2.43}$$

From above bases, we compute other set of projectors

$$\mathbf{Z}_{p_0} = \mathbf{p}_0^{*T}\mathbf{P}_1\mathbf{p}_0 = \begin{pmatrix} 0 & 0 \\ 0 & 1 \end{pmatrix}, \quad \text{and} \quad \mathbf{Z}_{q_0} = \mathbf{p}_0^{*T}\mathbf{Q}_1\mathbf{p}_0 = \begin{pmatrix} 1 & 0 \\ 0 & 0 \end{pmatrix}.$$

Next, we construct the basis matrices of these projectors and there respective inverses given by

$$\mathbf{z}_{p_0} = \begin{pmatrix} 0 \\ 1 \end{pmatrix}, \quad \mathbf{z}_{q_0} = \begin{pmatrix} 1 \\ 0 \end{pmatrix}, \quad \text{and} \quad \mathbf{z}_{p_0}^* = \begin{pmatrix} 0 \\ 1 \end{pmatrix}, \quad \mathbf{z}_{q_0}^* = \begin{pmatrix} 1 \\ 0 \end{pmatrix}. \tag{3.2.44}$$

Setting $\mu = 2$ and substituting Eqs. (3.2.43)–(3.2.44) into (3.2.10) leads to a completely decoupled system of (3.2.33) given by

$$\xi_p' = -\frac{1}{GL}\xi_p - \frac{1}{L}\mathbf{u}, \quad \xi_p(0) = \imath_L(0), \tag{3.2.45a}$$

$$\xi_{q,1} = 0\xi_p - \mathbf{u} \tag{3.2.45b}$$

$$\xi_{q,0} = \begin{pmatrix} 0 \\ 0 \end{pmatrix}\xi_p + \begin{pmatrix} -1 \\ 0 \end{pmatrix}\mathbf{u} + \begin{pmatrix} 0 \\ C \end{pmatrix}\xi_{q,1}', \tag{3.2.45c}$$

$$\mathbf{y} = -\xi_p + 0\xi_{q,1} + \begin{pmatrix} 0 & -1 \end{pmatrix}\xi_{q,0} \tag{3.2.45d}$$

We can observe that there is a completely decoupling between the differential and algebraic part since $\mathbf{A}_q = 0$. If we compare system (3.2.41) and (3.2.45), both lead to the same output solution given by $\mathbf{y} = -\xi_p + C\mathbf{u}'$.

Example 3.2.3 Consider a simple RL electric network in Fig. 3.2. We need to find the unknowns $x = [e_1, e_2, \imath_L]^T$ in the electric network. Using the Modified Nodal Analysis on this network leads to DAE system of the form (2.3.1), where $u = \imath(t)$ with system matrices given by

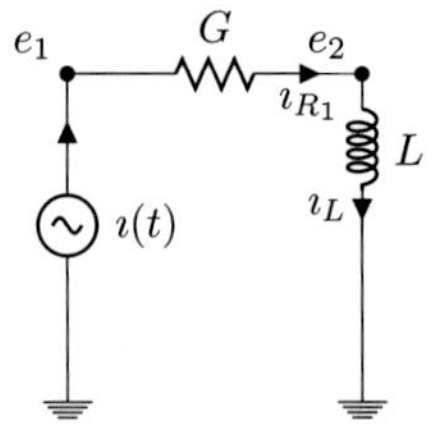

Fig. 3.2 Simple RL network

$$\mathbf{E} = \begin{pmatrix} 0 & 0 & 0 \\ 0 & 0 & 0 \\ 0 & 0 & L \end{pmatrix}, \quad \mathbf{A} = \begin{pmatrix} -G & G & 0 \\ G & -G & -1 \\ 0 & 1 & 0 \end{pmatrix}, \quad \mathbf{B} = \begin{pmatrix} 1 \\ 0 \\ 0 \end{pmatrix}, \quad \mathbf{C} = \mathbf{B}. \tag{3.2.46}$$

It is easy to check that this DAE has a regular matrix pencil, since $\det(\lambda E - A) = G > 0$ and its matrix pencil $(\mathbf{E}, \mathbf{A})$ has only infinite spectrum. Thus its decoupled system must be purely algebraic system and take the form (3.2.21). We can decouple DAE (3.2.46) following procedure in Sect. 3.2.2. First, we need to compute the matrix and projector chains as follows. This is done by combining Definition 3.1.1 and Theorem 3.1.1. We observed that this is an index-2 DAE. We can then choose the admissible projectors given by

$$\mathbf{Q}_0 = \begin{pmatrix} 1 & 0 & 0 \\ 0 & 1 & 0 \\ 0 & 0 & 0 \end{pmatrix}, \text{ and, } \mathbf{Q}_1 = \begin{pmatrix} 0 & 0 & L \\ 0 & 0 & L \\ 0 & 0 & 1 \end{pmatrix}, \tag{3.2.47}$$

such that $\mathbf{Q}_1\mathbf{Q}_0 = 0$. Then we have,

$$\mathbf{E}_2 = \begin{pmatrix} G & -G & 0 \\ -G & G & 1 \\ 0 & -1 & L \end{pmatrix}.$$

Since the $\mathbf{E}_2$ is non-singular, thus (3.2.46) is an index 2 DAE. Then, the basis chain and their respective left inverses are given by

$$\mathbf{p}_0 = \begin{pmatrix} 0 \\ 0 \\ 1 \end{pmatrix}, \quad \mathbf{q}_0 = \begin{pmatrix} 1 & 0 \\ 0 & 1 \\ 0 & 0 \end{pmatrix}, \quad \mathbf{p}_0{*}^T = \begin{pmatrix} 0 & 0 & 1 \end{pmatrix}, \quad \mathbf{q}_0{*}^T = \begin{pmatrix} 1 & 0 & 0 \\ 0 & 1 & 0 \end{pmatrix}. \tag{3.2.48}$$

Setting $\mu = 2$ and substituting Eqs. (3.2.47) and (3.2.48) into (3.2.21), we obtain,

$$\begin{aligned} \xi_{q,1} &= \mathbf{u} \\ \xi_{q,0} &= \begin{pmatrix} G^{-1} \\ 0 \end{pmatrix} \mathbf{u} + \begin{pmatrix} L \\ L \end{pmatrix} \xi'_{q,1}(t), \\ \mathbf{y} &= 0\xi_{q,1} + \begin{pmatrix} 1 & 0 \end{pmatrix} \xi_{q,0}. \end{aligned} \tag{3.2.49}$$

We can observe that system (3.2.49) can be computed explicitly without using any numerical integration techniques. Thus instead of solving DAE system (3.2.46) is easier to solve (3.2.49) and their solutions coincides given by $\mathbf{y} = G^{-1}\mathbf{u} + L\mathbf{u}'$.

Example 3.2.4 In this example, we consider a simple mechanical system whose linearized nonlinear equations of motion can be written in the form (2.3.1) with system matrices

$$
\mathbf{E} = \operatorname{diag}(1, 1, 1, m_1, m_2, m_2, 0), \quad \mathbf{A} = \begin{pmatrix} 0 & 0 & 0 & 1 & 0 & 0 & 0 \\ 0 & 0 & 0 & 0 & 1 & 0 & 0 \\ 0 & 0 & 0 & 0 & 0 & 1 & 0 \\ -m_2 g/L & m_2 g/L & 0 & 0 & 0 & 0 & 0 \\ m_2 g/L & -m_2 g/L & 0 & 0 & 0 & 0 & 0 \\ 0 & 0 & m_2 g/L & 0 & 0 & 0 & 2L \\ 0 & 0 & -2L & 0 & 0 & 0 & 0 \end{pmatrix},
$$

$$
\mathbf{B} = \begin{bmatrix} 0, 0, 0, 1, 0, 0, 0 \end{bmatrix}^T, \quad \mathbf{C} = \begin{pmatrix} 0 & 1 & 0 & 0 & 0 & 0 & 0 \\ 0 & 0 & 1 & 0 & 0 & 0 & 0 \end{pmatrix}^T, \tag{3.2.50}
$$

This system has a regular matrix pencil $(\mathbf{E}, \mathbf{A})$ and has the finite eigenvalues $\sigma_f(\mathbf{E}, \mathbf{A}) = \{0, 0, \pm i\sqrt{(m_1 + m_2)g/(m_1 L)}\}$. Hence, its decoupled system will take the form (3.2.10). First, we need to construct the matrix and projector chains. We observed that this is an index-3 system whose admissible projectors are given by

$$
\mathbf{Q}_0 = \begin{pmatrix} 0 & 0 & 0 & 0 & 0 & 0 & 0 \\ 0 & 0 & 0 & 0 & 0 & 0 & 0 \\ 0 & 0 & 0 & 0 & 0 & 0 & 0 \\ 0 & 0 & 0 & 0 & 0 & 0 & 0 \\ 0 & 0 & 0 & 0 & 0 & 0 & 0 \\ 0 & 0 & 0 & 0 & 0 & 0 & 0 \\ 0 & 0 & \frac{m_2}{2L} & 0 & 0 & \frac{m_2}{2L} & 1 \end{pmatrix}, \quad \mathbf{Q}_1 = \begin{pmatrix} 0 & 0 & 0 & 0 & 0 & 0 & 0 \\ 0 & 0 & 0 & 0 & 0 & 0 & 0 \\ 0 & 0 & 0 & 0 & 0 & 0 & 0 \\ 0 & 0 & 0 & 0 & 0 & 0 & 0 \\ 0 & 0 & 0 & 0 & 0 & 0 & 0 \\ 0 & 0 & 1 & 0 & 0 & 1 & 0 \\ 0 & 0 & 0 & 0 & 0 & 0 & 0 \end{pmatrix}, \quad \mathbf{Q}_2 = \begin{pmatrix} 0 & 0 & 0 & 0 & 0 & 0 & 0 \\ 0 & 0 & 0 & 0 & 0 & 0 & 0 \\ 0 & 0 & 1 & 0 & 0 & 0 & 0 \\ 0 & 0 & 0 & 0 & 0 & 0 & 0 \\ 0 & 0 & 0 & 0 & 0 & 0 & 0 \\ 0 & 0 & 0 & 0 & 0 & 0 & 0 \\ 0 & 0 & 0 & 0 & 0 & 0 & 0 \end{pmatrix} \tag{3.2.51}
$$

and final matrices on the matrix chain are given by,

$$
\mathbf{E}_3 = \begin{pmatrix} 1 & 0 & 0 & 0 & 0 & 0 & 0 \\ 0 & 1 & 0 & 0 & 0 & 0 & 0 \\ 0 & 0 & 1 & 0 & 0 & -1 & 0 \\ 0 & 0 & 0 & m_1 & 0 & 0 & 0 \\ 0 & 0 & 0 & 0 & m_2 & 0 & 0 \\ 0 & 0 & \frac{gm_2}{L} & 0 & 0 & m_2 & -2L \\ 0 & 0 & 2L & 0 & 0 & 0 & 0 \end{pmatrix}, \quad \mathbf{A}_3 = \begin{pmatrix} 0 & 0 & 0 & 1 & 0 & 0 & 0 \\ 0 & 0 & 0 & 0 & 1 & 0 & 0 \\ 0 & 0 & 0 & 0 & 0 & 0 & 0 \\ -\frac{gm_2}{L} & \frac{gm_2}{L} & 0 & 0 & 0 & 0 & 0 \\ \frac{gm_2}{L} & -\frac{gm_2}{L} & 0 & 0 & 0 & 0 & 0 \\ 0 & 0 & 0 & 0 & 0 & 0 & 0 \\ 0 & 0 & 0 & 0 & 0 & 0 & 0 \end{pmatrix}. \tag{3.2.52}
$$

Since $\mathbf{E}_3$ is non-singular, then this system is of tractability index-3 or index-3 DAE. Then, we constructed the basis vector $(\boldsymbol{p}_0, \boldsymbol{q}_0)$ and their corresponding left inverses given by

$$
\boldsymbol{p}_0 = \begin{pmatrix} 1 & 0 & 0 & 0 & 0 & 0 \\ 0 & 1 & 0 & 0 & 0 & 0 \\ 0 & 0 & 0 & 0 & -1 & -\frac{2L}{m_2} \\ 0 & 0 & 1 & 0 & 0 & 0 \\ 0 & 0 & 0 & 1 & 0 & 0 \\ 0 & 0 & 0 & 0 & 1 & 0 \\ 0 & 0 & 0 & 0 & 0 & 1 \end{pmatrix}, \ \boldsymbol{q}_0 = \begin{pmatrix} 0 \\ 0 \\ 0 \\ 0 \\ 0 \\ 0 \\ 1 \end{pmatrix},
$$

$$
\boldsymbol{p}_0^{*T} = \begin{pmatrix} 1 & 0 & 0 & 0 & 0 & 0 & 0 \\ 0 & 1 & 0 & 0 & 0 & 0 & 0 \\ 0 & 0 & 0 & 1 & 0 & 0 & 0 \\ 0 & 0 & 0 & 0 & 1 & 0 & 0 \\ 0 & 0 & 0 & 0 & 0 & 1 & 0 \\ 0 & 0 & -\frac{m_2}{2L} & 0 & 0 & -\frac{m_2}{2L} & 0 \end{pmatrix}, \quad \boldsymbol{q}_0^{*T} = \begin{pmatrix} 0 \\ 0 \\ \frac{m_2}{2L} \\ 0 \\ 0 \\ \frac{m_2}{2L} \\ 1 \end{pmatrix}, \tag{3.2.53}
$$

for the projector $\mathbf{Q}_0$ and its complementary $\mathbf{P}_0$, respectively. Then, we use the above basis to construct the second basis $(\mathbf{z}_{p_0}, \mathbf{z}_{q_0})$ and their corresponding left inverses $\begin{pmatrix} \mathbf{z}_{p_0}^{*T} \\ \mathbf{z}_{q_0}^{*T} \end{pmatrix}$ given by,

$$
\mathbf{z}_{p_0} = \begin{pmatrix} 1 & 0 & 0 & 0 & 0 \\ 0 & 1 & 0 & 0 & 0 \\ 0 & 0 & 1 & 0 & 0 \\ 0 & 0 & 0 & 1 & 0 \\ 0 & 0 & 0 & 0 & 1 \\ 0 & 0 & 0 & 0 & 0 \end{pmatrix}, \ \mathbf{z}_{q_0} = \begin{pmatrix} 0 \\ 0 \\ 0 \\ 0 \\ -\frac{2L}{m_2} \\ 1 \end{pmatrix}, \ \mathbf{z}_{p_0}^{*T} = \begin{pmatrix} 1 & 0 & 0 & 0 & 0 & 0 \\ 0 & 1 & 0 & 0 & 0 & 0 \\ 0 & 0 & 1 & 0 & 0 & 0 \\ 0 & 0 & 0 & 1 & 0 & 0 \\ 0 & 0 & 0 & 0 & 1 & \frac{2L}{m_2} \end{pmatrix}, \ \mathbf{z}_{q_0}^{*T} = \begin{pmatrix} 0 \\ 0 \\ 0 \\ 0 \\ 0 \\ 1 \end{pmatrix}^T, \tag{3.2.54}
$$

for the projector $\mathbf{Z}_{p_0}$ and its complementary $\mathbf{Z}_{q_0}$, respectively. Thus, we use the above bases to construct the third basis $(\mathbf{z}_{p_1}, \mathbf{z}_{q_1})$ and their corresponding left inverses $\begin{pmatrix} \mathbf{z}_{p_1}^{*T} \\ \mathbf{z}_{q_1}^{*T} \end{pmatrix}$ given by,

$$
\mathbf{z}_{p_1} = \begin{pmatrix} 1 & 0 & 0 & 0 \\ 0 & 1 & 0 & 0 \\ 0 & 0 & 1 & 0 \\ 0 & 0 & 0 & 1 \\ 0 & 0 & 0 & 0 \end{pmatrix}, \ \mathbf{z}_{q_1} = \begin{pmatrix} 0 \\ 0 \\ 0 \\ 0 \\ 1 \end{pmatrix}, \ \mathbf{z}_{p_1}^{*T} = \begin{pmatrix} 1 & 0 & 0 & 0 & 0 \\ 0 & 1 & 0 & 0 & 0 \\ 0 & 0 & 1 & 0 & 0 \\ 0 & 0 & 0 & 1 & 0 \end{pmatrix}, \ \mathbf{z}_{q_1}^{*T} = \begin{pmatrix} 0 & 0 & 0 & 0 & 1 \end{pmatrix}, \tag{3.2.55}
$$

for the projector $\mathbf{Z}_{p_1}$ and its complementary $\mathbf{Z}_{q_1}$, respectively. Substituting $\mu = 3$ and (3.2.53)–(3.2.55) into (3.2.10), we obtain the decoupled system of (3.2.50) given by

$$
\begin{aligned}
\xi_p' &= \mathbf{A}_p\xi_p + \mathbf{B}_p\mathbf{u},\\
\xi_{q,2} &= \mathbf{A}_{q,2}\xi_p + \mathbf{B}_{q,2}\mathbf{u},\\
\xi_{q,1} &= \mathbf{A}_{q,1}\xi_p + \mathbf{B}_{q,1}\mathbf{u} + \mathbf{A}_{q_{1,2}}\xi_{q,2}',\\
\xi_{q,0} &= \mathbf{A}_{q,0}\xi_p + \mathbf{B}_{q,0}\mathbf{u} + \mathbf{A}_{q_{0,1}}\xi_{q,1}' + \mathbf{A}_{q_{0,2}}\xi_{q,2}',\\
\mathbf{y} &= \mathbf{C}_p^T\xi_p + \mathbf{C}_{q,2}^T\xi_{q,2} + \mathbf{C}_{q,1}^T\xi_{q,1} + \mathbf{C}_{q,0}^T\xi_{q,0},
\end{aligned}
$$

with matrix coefficients,

$$
\mathbf{A}_p = \begin{pmatrix} 0 & 0 & 1 & 0\\ 0 & 0 & 0 & 1\\ -m_2g/(Lm_1) & m_2g/(Lm_1) & 0 & 0\\ g/L & -g/L & 0 & 0 \end{pmatrix}, \quad \mathbf{B}_p = \begin{pmatrix} 0\\ 0\\ 1/m_1\\ 0 \end{pmatrix},
$$

$$
\begin{aligned}
&\mathbf{A}_{q,2} = \begin{pmatrix} 0 & 0 & 0 \end{pmatrix}, \quad \mathbf{B}_{q,2} = 0,\\
&\mathbf{A}_{q,1} = \begin{pmatrix} 0 & 0 & 0 & 0 \end{pmatrix}, \quad \mathbf{B}_{q,1} = 0,\\
&\mathbf{A}_{q_{1,2}} = \frac{m_2}{2L}, \quad \mathbf{A}_{q,0} = \begin{pmatrix} 0 & 0 & 0 & 0 \end{pmatrix}, \quad \mathbf{B}_{q,0} = 0, \quad \mathbf{A}_{q_{0,1}} = -1,\\
&\mathbf{A}_{q_{0,2}} = 0, \quad \mathbf{C}_p = \begin{pmatrix} 0 & 1 & 0 & 0\\ 0 & 0 & 0 & 0 \end{pmatrix}^T, \quad \mathbf{C}_{q,2} = \begin{pmatrix} 0\\ 1 \end{pmatrix}^T,\\
&\mathbf{C}_{q,1} = \begin{pmatrix} 0\\ 0 \end{pmatrix}^T, \quad \mathbf{C}_{q,0} = \begin{pmatrix} 0\\ 0 \end{pmatrix}^T.
\end{aligned}
$$

We can observe that this is system is completely decoupled and can be reduced to an ODE of dimension 4 given by

$$
\begin{aligned}
\xi_p' &= \begin{pmatrix} 0 & 0 & 1 & 0\\ 0 & 0 & 0 & 1\\ -m_2g/(Lm_1) & m_2g/(Lm_1) & 0 & 0\\ g/L & -g/L & 0 & 0 \end{pmatrix}\xi_p + \begin{pmatrix} 0\\ 0\\ 1/m_1\\ 0 \end{pmatrix}\mathbf{u},\\
\mathbf{y} &= \begin{pmatrix} 0 & 1 & 0 & 0\\ 0 & 0 & 0 & 0 \end{pmatrix}\xi_p.
\end{aligned} \tag{3.2.56}
$$

Example 3.2.5 Consider an index-3 DAE obtained from [33] with system matrices given by

$$
\mathbf{E} = \begin{pmatrix} 0 & 1 & 0\\ 0 & 0 & 1\\ 0 & 0 & 0 \end{pmatrix}, \quad \mathbf{A} = \begin{pmatrix} 1 & 0 & 0\\ 0 & 1 & 0\\ 0 & 0 & 1 \end{pmatrix}, \quad \mathbf{B} = \begin{pmatrix} 10\\ 0.1\\ 0 \end{pmatrix}, \quad \mathbf{C} = \begin{pmatrix} 0.04\\ 30\\ 1 \end{pmatrix}. \tag{3.2.57}
$$

Since the $\det(\lambda\mathbf{E} - \mathbf{A}) = -1$, thus the DAE system (3.2.57) is solvable and its matrix pencil has no finite eigenvalues. Hence we expect its decoupled system to have no differential part and takes the form (3.2.21). In order to decouple this DAE system, we choose the admissible projector chains,

$$\mathbf{Q}_0 = \begin{pmatrix} 1 & 1 & 1 \\ 0 & 0 & 0 \\ 0 & 0 & 0 \end{pmatrix}, \; \mathbf{Q}_1 = \begin{pmatrix} 0 & 0 & 0 \\ 0 & 1 & 1 \\ 0 & 0 & 0 \end{pmatrix}, \; \mathbf{Q}_2 = \begin{pmatrix} 0 & 0 & 0 \\ 0 & 0 & 0 \\ 0 & 0 & 1 \end{pmatrix}$$

that satisfy the condition $\mathbf{Q}_j\mathbf{Q}_i = 0$, $j > i$, $i, j = 0, 1, 2$ and the last matrix chain is given by $\mathbf{E}_3 = \begin{pmatrix} -1 & 1 & 0 \\ 0 & -1 & 1 \\ 0 & 0 & -1 \end{pmatrix}$. Following the procedure discussed in the beginning of this subsection, we were able to construct projector basis chain and there respective left inverses given by

$$\boldsymbol{p}_0 = \begin{pmatrix} -1 & -1 \\ 1 & 0 \\ 0 & 1 \end{pmatrix}, \quad \boldsymbol{q}_0 = \begin{pmatrix} 1 \\ 0 \\ 0 \end{pmatrix}, \quad \boldsymbol{p}_0^{*T} = \begin{pmatrix} 0 & 1 & 0 \\ 0 & 0 & 1 \end{pmatrix}, \quad \boldsymbol{q}_0^{*T} = \begin{pmatrix} 1 & 1 & 1 \end{pmatrix},$$

$$\text{and} \quad \boldsymbol{z}_{\boldsymbol{p}_0} = \begin{pmatrix} -1 \\ 1 \end{pmatrix}, \quad \boldsymbol{z}_{\boldsymbol{q}_0} = \begin{pmatrix} 1 \\ 0 \end{pmatrix}, \quad \boldsymbol{z}_{\boldsymbol{p}_0}^{*T} = \begin{pmatrix} 0 & 1 \end{pmatrix}, \quad \boldsymbol{z}_{\boldsymbol{q}_0}^{*T} = \begin{pmatrix} 1 & 1 \end{pmatrix}.$$

Thus, substituting $\mu = 3$ and the above matrices into (3.2.21), we obtain the a decoupled system of (3.2.57) given by

$$\begin{aligned} \xi_{q,2} &= 0\boldsymbol{u}, \\ \xi_{q,1} &= -0.1\boldsymbol{u} + \xi'_{q,2}, \\ \xi_{q,0} &= -10.1\boldsymbol{u} + \xi'_{q,1} + 0\xi'_{q,2}, \\ \mathbf{y} &= -29\xi_{q,2} + 29.96\xi_{q,1} + 0.04\xi_{q,0}. \end{aligned} \tag{3.2.58}$$

We can observe that it is easy to solve (3.2.58) and its solution coincide with that of (3.2.57) given by $\boldsymbol{y} = -3.4\boldsymbol{u} - 0.004\boldsymbol{u}'$.

3.3 Implicit Decoupling of DAEs Using Matrix, Projector and Basis Chains

In the previous, section we have discussed the decoupling of DAEs using matrix, projector and basis chains. However, this approach has an inherited limitation of inversion of matrix $\mathbf{E}_\mu$ from (3.1.4) which can be computationally expensive for large-scale matrices. In this section, we intend to decouple DAEs without inverting matrix $\mathbf{E}_\mu$. This implies that (3.1.3) is our starting equation.

3.4 Index-1 DAEs

If, we consider index-1 DAEs, i.e., $\mu = 1$. Then, substituting $\mu = 1$ into (3.1.3), we obtain:

$$\mathbf{E}_1\Big[\mathbf{P}_0\dot{\boldsymbol{x}} + \mathbf{Q}_0\boldsymbol{x}\Big] = \mathbf{A}_1\boldsymbol{x} + \mathbf{B}\boldsymbol{u}. \tag{3.4.1}$$

Using $\boldsymbol{p}_0 \in \operatorname{Im}\mathbf{P}_0$ and $\boldsymbol{q}_0 \in \operatorname{Ker}\mathbf{E}_0$ as defined in Sect. 3.2.1, then we can decompose $\boldsymbol{x}$ as

$$\boldsymbol{x} = \begin{pmatrix}\boldsymbol{p}_0 & \boldsymbol{q}_0\end{pmatrix}\begin{pmatrix}\xi_p\\ \xi_q\end{pmatrix}, \tag{3.4.2}$$

where $\xi_p \in \mathbb{R}^{n_0}$ and $\xi_q \in \mathbb{R}^{k_0}$ are the projected differential and algebraic variables, respectively. Substituting (3.4.2) into (3.4.1), we obtain

$$\begin{pmatrix}\mathbf{E}_1\boldsymbol{p}_0 & 0\end{pmatrix}\begin{pmatrix}\xi_p\\ \xi_q\end{pmatrix}' = \begin{pmatrix}\mathbf{A}_1\boldsymbol{p}_0 & -\mathbf{E}_1\boldsymbol{q}_0\end{pmatrix}\begin{pmatrix}\xi_p\\ \xi_q\end{pmatrix} + \mathbf{B}\boldsymbol{u}. \tag{3.4.3}$$

Next, we need to construct other bases $\hat{\boldsymbol{p}}_0$ and $\hat{\boldsymbol{q}}_0$ with the same dimension as $\boldsymbol{p}_0$ and $\boldsymbol{q}_0$, respectively. This is done as follows. Left multiplying (3.4.3) by $\begin{pmatrix}\hat{\boldsymbol{p}}_0 & \hat{\boldsymbol{q}}_0\end{pmatrix}^T \in \mathbb{R}^{n\times n}$ leads to

$$\begin{pmatrix}\hat{\boldsymbol{p}}_0^T\mathbf{E}_1\boldsymbol{p}_0 & 0\\ \hat{\boldsymbol{q}}_0^T\mathbf{E}_1\boldsymbol{p}_0 & 0\end{pmatrix}\begin{pmatrix}\xi_p\\ \xi_q\end{pmatrix}' = \begin{pmatrix}\hat{\boldsymbol{p}}_0^T\mathbf{A}_1\boldsymbol{p}_0 & -\hat{\boldsymbol{p}}_0^T\mathbf{E}_1\boldsymbol{q}_0\\ \hat{\boldsymbol{q}}_0^T\mathbf{A}_1\boldsymbol{p}_0 & -\hat{\boldsymbol{q}}_0^T\mathbf{E}_1\boldsymbol{q}_0\end{pmatrix}\begin{pmatrix}\xi_p\\ \xi_q\end{pmatrix} + \begin{pmatrix}\hat{\boldsymbol{p}}_0^T\mathbf{B}\\ \hat{\boldsymbol{q}}_0^T\mathbf{B}\end{pmatrix}\mathbf{u}. \tag{3.4.4}$$

From (3.4.4), we can observe that we still have a cross coupling between differential and algebraic parts and it not also clear how we can construct bases $\hat{\boldsymbol{p}}_0$ and $\hat{\boldsymbol{q}}_0$. We can observe that if $\hat{\boldsymbol{p}}_0 \in \operatorname{Ker}\boldsymbol{q}_0^T\mathbf{E}_1^T$ and $\hat{\boldsymbol{q}}_0 \in \operatorname{Ker}\boldsymbol{p}_0^T\mathbf{E}_1^T$, then this cross coupling can be eliminated. We can easily check it is the same as $\hat{\boldsymbol{p}}_0 \in \operatorname{Ker}\boldsymbol{q}_0^T\mathbf{A}^T$ and $\hat{\boldsymbol{q}}_0 \in \operatorname{Ker}\mathbf{E}^T$. This implies, $\hat{\boldsymbol{p}}_0^T\mathbf{E}_1\boldsymbol{q}_0 = \mathbf{0}$ and $\hat{\boldsymbol{q}}_0^T\mathbf{E}_1\boldsymbol{p}_0 = \mathbf{0}$. Thus, (3.4.4), simplifies to

$$\begin{pmatrix}\hat{\boldsymbol{p}}_0^T\mathbf{E}_1\boldsymbol{p}_0 & \mathbf{0}\\ \mathbf{0} & \mathbf{0}\end{pmatrix}\begin{pmatrix}\xi_p\\ \xi_q\end{pmatrix}' = \begin{pmatrix}\hat{\boldsymbol{p}}_0^T\mathbf{A}_1\boldsymbol{p}_0 & \mathbf{0}\\ \hat{\boldsymbol{q}}_0^T\mathbf{A}_1\boldsymbol{p}_0 & -\hat{\boldsymbol{q}}_0^T\mathbf{E}_1\boldsymbol{q}_0\end{pmatrix}\begin{pmatrix}\xi_p\\ \xi_q\end{pmatrix} + \begin{pmatrix}\hat{\boldsymbol{p}}_0^T\mathbf{B}\\ \hat{\boldsymbol{q}}_0^T\mathbf{B}\end{pmatrix}\mathbf{u}. \tag{3.4.5}$$

Using the fact that $\mathbf{E}_1 = \mathbf{E} - \mathbf{A}\mathbf{Q}_0$ and $\mathbf{A}_1 = \mathbf{A}\mathbf{P}_0$, then the (3.4.5) simplifies to an equivalent implicit decoupled system of (2.3.1) given by

$$\mathbf{E}_p\xi_p' = \mathbf{A}_p\xi_p + \mathbf{B}_p\boldsymbol{u}, \tag{3.4.6a}$$

$$\mathbf{E}_q\xi_q = \mathbf{A}_q\xi_p + \mathbf{B}_q\boldsymbol{u}, \tag{3.4.6b}$$

$$\mathbf{y} = \mathbf{C}_p^T\xi_p + \mathbf{C}_q^T\xi_q, \tag{3.4.6c}$$

where $\mathbf{E}_p = \hat{\boldsymbol{p}}_0^T \mathbf{E} \boldsymbol{p}_0$, $\mathbf{A}_p = \hat{\boldsymbol{p}}_0^T \mathbf{A} \boldsymbol{p}_0 \in \mathbb{R}^{n_p \times n_p}$, $\mathbf{B}_p = \hat{\boldsymbol{p}}_0^T \mathbf{B} \in \mathbb{R}^{n_p \times m}$, $\mathbf{E}_q = -\hat{\boldsymbol{q}}_0^T \mathbf{A} \boldsymbol{q}_0 \in \mathbb{R}^{n_q \times n_q}$, $\mathbf{A}_q = \hat{\boldsymbol{q}}_0^T \mathbf{A} \boldsymbol{p}_0 \in \mathbb{R}^{n_p \times n_q}$, $\mathbf{B}_q = \hat{\boldsymbol{q}}_0^T \mathbf{B} \in \mathbb{R}^{n_q \times m}$ and $\mathbf{C}_p = \boldsymbol{p}^T \mathbf{C} \in \mathbb{R}^{n_p, \ell}$, $\mathbf{C}_q = \boldsymbol{q}^T \mathbf{C} \in \mathbb{R}^{n_q, \ell}$. We note that matrices $\mathbf{E}_p$ and $\mathbf{E}_q$ are always nonsingular.

3.5 Index-2 DAEs

Here, we consider index-2 DAEs, i.e., $\mu = 2$. Then, substituting $\mu = 2$ into (3.1.3), we obtain:

$$\mathbf{E}_2 \Big[\mathbf{P}_1 \mathbf{P}_0 \dot{\boldsymbol{x}} + \mathbf{Q}_0 \boldsymbol{x} + \mathbf{Q}_1 \boldsymbol{x} \Big] = \mathbf{A}_2 \boldsymbol{x} + \mathbf{B} \boldsymbol{u}, \tag{3.5.1}$$

In the previous section, we discussed that for higher index DAEs there is a possibility of obtaining a decoupled system with either a differential part or without a differential part depending on the nature of the spectrum of the matrix pencil. Thus for the case of index-2 DAEs, we shall consider two cases as follows:

3.5.1 Index-2 DAEs with a Finite Spectrum

Assume that the matrix pencil $(\mathbf{E}, \mathbf{A})$ of (3.2.48) has at least one finite eigenvalue. Then, its state-space vector $\boldsymbol{x}$ can be decomposed as

$$\boldsymbol{x} = \begin{pmatrix} \boldsymbol{p} z_{\boldsymbol{p}_0} & \boldsymbol{p} z_{\boldsymbol{q}_0} & \boldsymbol{q} \end{pmatrix} \begin{pmatrix} \xi_p \\ \xi_{q,1} \\ \xi_{q,0} \end{pmatrix} \in \mathbb{R}^n, \tag{3.5.2}$$

where $\xi_p \in \mathbb{R}^{n_p}$, $\xi_{q,1} \in \mathbb{R}^{k_1}$, $\xi_{q,0} \in \mathbb{R}^{k_0}$ and $n = n_p + k_1 + k_0$. $\mathbf{z}_{\boldsymbol{p}_0}$ and $\mathbf{z}_{\boldsymbol{q}_0}$ are as defined in Sect. 3.2.1. Substituting (3.5.2) into (3.5.1), we obtain,

$$\begin{pmatrix} \mathbf{E}_2 \boldsymbol{p} z_{\boldsymbol{p}_0} & -\mathbf{E}_2 \mathbf{Q}_0 \mathbf{Q}_1 \boldsymbol{p} z_{\boldsymbol{q}_0} & 0 \end{pmatrix} \begin{pmatrix} \xi_p \\ \xi_{q,1} \\ \xi_{q,0} \end{pmatrix}' = \begin{pmatrix} \mathbf{A}_2 \boldsymbol{p} z_{\boldsymbol{p}_0} & -\mathbf{E}_2 \mathbf{Q}_1 \boldsymbol{p} z_{\boldsymbol{q}_0} & -\mathbf{E}_2 \boldsymbol{q} \end{pmatrix} \begin{pmatrix} \xi_p \\ \xi_{q,1} \\ \xi_{q,0} \end{pmatrix} + \mathbf{B} \boldsymbol{u}. \tag{3.5.3}$$

Using the same trick as in the previous case, we need to construct another set of bases $\hat{\boldsymbol{p}}_0^T \in \mathbb{R}^{n_0 \times n}$, $\hat{\boldsymbol{q}}_0^T \in \mathbb{R}^{k_0 \times n}$, $\hat{z}_{\boldsymbol{p}_0}^T \in \mathbb{R}^{k_1, n_0}$ and $\hat{z}_{\boldsymbol{q}_0}^T \in \mathbb{R}^{k_0 \times n_0}$ in order to obtain a one way coupling between the differential and algebraic parts. If we let $\hat{\boldsymbol{p}}_0 \in \operatorname{Ker} \boldsymbol{q}^T \mathbf{E}_2^T$, $\hat{\boldsymbol{q}}_0 \in \operatorname{Ker} \boldsymbol{p}^T \mathbf{E}_2^T$, $\hat{z}_{\boldsymbol{p}_0} \in \operatorname{Ker} (\hat{\boldsymbol{p}}_0^T \mathbf{E}_2 \boldsymbol{p} z_{\boldsymbol{q}_0})^T$ and $\hat{z}_{\boldsymbol{q}_0} = \operatorname{Ker} (\hat{\boldsymbol{p}}_0^T \mathbf{E}_2 \boldsymbol{p} z_{\boldsymbol{p}_0})^T$. Thus, multiplying (3.5.3) by $\begin{pmatrix} \hat{\boldsymbol{p}}_0 \hat{z}_{\boldsymbol{p}_0} & \hat{\boldsymbol{p}}_0 \hat{z}_{\boldsymbol{q}_0} & \hat{\boldsymbol{q}}_0 \end{pmatrix}^T$ and simplifying, we obtain

$$\begin{pmatrix} \hat{z}_{p_0}^T \hat{p}_0^T \mathbf{E}_2 p z_{p_0} & 0 & 0 \\ 0 & 0 & 0 \\ 0 & -\hat{q}_0^T \mathbf{E}_2 \mathbf{Q}_1 p z_{q_0} & 0 \end{pmatrix} \begin{pmatrix} \xi_p \\ \xi_{q,1} \\ \xi_{q,0} \end{pmatrix}'$$

$$= \begin{pmatrix} \hat{z}_{p_0}^T \hat{p}_0^T \mathbf{A}_2 p z_{p_0} & 0 & 0 \\ \hat{z}_{q_0}^T \hat{p}_0^T \mathbf{A}_2 p z_{p_0} & -\hat{z}_{q_0}^T \hat{p}_0^T \mathbf{E}_2 p z_{q_0} & 0 \\ \hat{q}_0^T \mathbf{A}_2 p z_{p_0} & -\hat{q}_0^T \mathbf{E}_2 \mathbf{Q}_1 p z_{q_0} & -\hat{q}_0^T \mathbf{E}_2 q \end{pmatrix} \begin{pmatrix} \xi_p \\ \xi_{q,1} \\ \xi_{q,0} \end{pmatrix} + \begin{pmatrix} \hat{z}_{p_0}^T \hat{p}_0^T \mathbf{B} \\ \hat{z}_{q_0}^T \hat{p}_0^T \mathbf{B} \\ \hat{q}_0^T \mathbf{B} \end{pmatrix} u.$$

From the above system, without loss of generality the implicit decoupled system of (2.3.1) is given by

$$\mathbf{E}_p \xi_p' = \mathbf{A}_p \xi_p + \mathbf{B}_p u, \tag{3.5.4a}$$

$$\mathbf{E}_{q,1} \xi_{q,1} = \mathbf{A}_{q,1} \xi_p + \mathbf{B}_{q,1} \mathbf{u}, \tag{3.5.4b}$$

$$\mathbf{E}_{q,0} \xi_{q,0} = \mathbf{A}_{q,0} \xi_p + \mathbf{B}_{q,0} \mathbf{u} + \mathbf{A}_{q_{0,1}} \Big[\xi_{q,1}' - \xi_{q,1} \Big], \tag{3.5.4c}$$

$$\mathbf{y} = \mathbf{C}_p^T \xi_p + \mathbf{C}_{q,1}^T \xi_{q,1} + \mathbf{C}_{q,0}^T \xi_{q,0} \tag{3.5.4d}$$

where

$\mathbf{E}_p = \hat{z}_{p_0}^T \hat{p}_0^T \mathbf{E} p z_{p_0} \in \mathbb{R}^{n_p \times n_p}$, $\mathbf{A}_p = \hat{z}_{p_0}^T \hat{p}_0^T \mathbf{A} p z_{p_0} \in \mathbb{R}^{n_p \times n_p}$, $\mathbf{B}_p = \hat{z}_{p_0}^T \hat{p}_0^T \mathbf{B} \in \mathbb{R}^{n_p \times m}$,
$\mathbf{E}_{q,1} = \hat{z}_{q_0}^T \hat{p}_0^T \mathbf{E}_2 p z_{q_0} \in \mathbb{R}^{k_1 \times k_1}$ $\mathbf{A}_{q,1} = \hat{z}_{q_0}^T \hat{p}_0^T \mathbf{A} p z_{p_0} \in \mathbb{R}^{k_1 \times n_p}$, $\mathbf{B}_{q,1} = \hat{z}_{q_0}^T \hat{p}_0^T \mathbf{B} \in \mathbb{R}^{k_1 \times m}$,
$\mathbf{E}_{q,0} = -\hat{q}_0^T \mathbf{A} q_0 \in \mathbb{R}^{k_0 \times k_0}$, $\mathbf{A}_{q,0} = \hat{q}_0^T \mathbf{A} p z_{p_0} \in \mathbb{R}^{k_0 \times n_p}$ $\mathbf{B}_{q,0} = \hat{q}_0^T \mathbf{B} \in \mathbb{R}^{k_0 \times m}$,
$\mathbf{A}_{q_{0,1}} = -\hat{q}_0^T \mathbf{A} p z_{q_0} \in \mathbb{R}^{k_0 \times k_1}$, $\mathbf{C}_p = z_{p_0}^T p^T \mathbf{C} \in \mathbb{R}^{n_p \times \ell}$, $\mathbf{C}_{q,1} = z_{q_0}^T p^T \mathbf{C} \in \mathbb{R}^{k_1 \times \ell}$,
and $\mathbf{C}_{q,0} = q^T \mathbf{C} \in \mathbb{R}^{k_0 \times \ell}$.

We note that matrices $\mathbf{E}_p$, $\mathbf{E}_{q,1}$ and $\mathbf{E}_{q,0}$ are always nonsingular.

3.5.2 Index-2 DAEs with Only Infinite Spectrum

Here, we assume that the matrix pencil $(\mathbf{E}, \mathbf{A})$ of (2.3.1) has no finite eigenvalues, i.e., $\sigma_f(\mathbf{E}, \mathbf{A}) = \emptyset$. This implies that $\mathbf{P}_0\mathbf{P}_1 = 0$, thus (3.5.1) simplifies to,

$$\mathbf{E}_2 \Big[\mathbf{P}_1 \mathbf{P}_0 \dot{x} + \mathbf{Q}_0 x + \mathbf{Q}_1 x \Big] = \mathbf{B} \mathbf{u}. \tag{3.5.5}$$

Recall from Sect. 3.2.2 that the state-space of index-2 DAE with only infinite spectrum can decompose x as, $x = (p \ q) \begin{pmatrix} \xi_{q,1} \\ \xi_{q,0} \end{pmatrix}$, $\xi_{q,1} \in \mathbb{R}^{k_1}$, $\xi_{q,0} \in \mathbb{R}^{k_0}$. Substituting x into (3.5.5) and simplifying, we obtain:

$$(\mathbf{E}_2 \mathbf{P}_1 p_0 \ \ 0) \begin{pmatrix} \xi_{q,1} \\ \xi_{q,0} \end{pmatrix}' = - (\mathbf{E}_2 \mathbf{Q}_1 p_0 \ \ \mathbf{E}_2 q_0) \begin{pmatrix} \xi_{q,1} \\ \xi_{q,0} \end{pmatrix} + \mathbf{B} \mathbf{u}. \tag{3.5.6}$$

In order to decouple (3.5.6), we introduce $\hat{\boldsymbol{p}}_0 \in \operatorname{Ker} \boldsymbol{q}^T\mathbf{E}_2^T$, $\hat{\boldsymbol{q}}_0 \in \operatorname{Ker} \boldsymbol{p}^T\mathbf{Q}_1^T\mathbf{E}_2^T$. Next, multiplying (3.5.3) by $\begin{pmatrix}\hat{\boldsymbol{p}}_0 & \hat{\boldsymbol{q}}_0\end{pmatrix}^T$, we obtain:

$$\begin{pmatrix} \mathbf{0} & \mathbf{0} \\ \hat{\boldsymbol{q}}_0^T\mathbf{E}_2\mathbf{P}_1\boldsymbol{p}_0 & \mathbf{0} \end{pmatrix} \begin{pmatrix} \xi_{q,1} \\ \xi_{q,0} \end{pmatrix}' + \begin{pmatrix} \hat{\boldsymbol{p}}_0^T\mathbf{E}_2\mathbf{Q}_1\boldsymbol{p}_0 & \mathbf{0} \\ \mathbf{0} & \hat{\boldsymbol{q}}_0^T\mathbf{E}_2\boldsymbol{q}_0 \end{pmatrix} \begin{pmatrix} \xi_{q,1} \\ \xi_{q,0} \end{pmatrix} = \begin{pmatrix} \hat{\boldsymbol{p}}_0^T\mathbf{B} \\ \hat{\boldsymbol{q}}_0^T\mathbf{B} \end{pmatrix} \mathbf{u}.$$

Thus, if (2.3.1) is of index-2 with only infinite spectrum its equivalent implicit decoupled system can be written in the form

$$\begin{aligned} \mathbf{E}_{q,1}\xi_{q,1} &= \mathbf{B}_{q,1}\mathbf{u}, \\ \mathbf{E}_{q,0}\xi_{q,0} &= \mathbf{B}_{q,0}\mathbf{u} + \mathbf{A}_{\boldsymbol{q}_{0,1}}\xi_{q,1}', \\ \mathbf{y} &= \mathbf{C}_{q,1}^T\,\xi_{q,1} + \mathbf{C}_{q,0}^T\,\xi_{q,0}, \end{aligned} \tag{3.5.7}$$

where $\mathbf{E}_{q,1} = \hat{\boldsymbol{p}}_0^T\mathbf{E}_2\mathbf{Q}_1\boldsymbol{p}_0 \in \mathbb{R}^{k_1\times k_1}$, $\mathbf{B}_{q,1} = \hat{\boldsymbol{p}}_0^T\mathbf{B} \in \mathbb{R}^{k_1\times m}$, $\mathbf{E}_{q,0} = \hat{\boldsymbol{q}}_0^T\mathbf{E}_2\boldsymbol{q}_0 \in \mathbb{R}^{k_0\times k_0}$, $\mathbf{B}_{q,0} = \hat{\boldsymbol{q}}_0^T\mathbf{B} \in \mathbb{R}^{k_0\times m}$, $\mathbf{A}_{\boldsymbol{q}_{0,1}} = -\hat{\boldsymbol{q}}_0^T\mathbf{E}_2\mathbf{P}_1\boldsymbol{p} \in \mathbb{R}^{k_0\times k_1}$, $\mathbf{C}_{q,1} = \boldsymbol{p}_0^T\mathbf{C} \in \mathbb{R}^{k_1\times \ell}$ and $\mathbf{C}_{q,0} = \boldsymbol{q}_0^T\mathbf{C} \in \mathbb{R}^{k_0\times \ell}$.

3.6 Index-3 DAEs

Here, we consider index-3 DAEs, i.e., $\mu = 3$. Then, substituting $\mu = 3$ into (3.1.3), we obtain:

$$\mathbf{E}_3\Big[\mathbf{P}_2\mathbf{P}_1\mathbf{P}_0\boldsymbol{x}' + \mathbf{Q}_2\boldsymbol{x} + \mathbf{Q}_1\boldsymbol{x} + \mathbf{Q}_0\boldsymbol{x}\Big] = \mathbf{A}_3\boldsymbol{x} + \mathbf{B}\mathbf{u}. \tag{3.6.1}$$

For the case of index-3 DAEs we also have two possibilities depending on the spectrum of the matrix pencil $(\mathbf{E}, \mathbf{A})$ as follows:

3.6.1 Index-3 DAEs with a Finite Spectrum

Assume that the matrix pencil $(\mathbf{E}, \mathbf{A})$ of (3.2.48) has at least one finite eigenvalues, i.e., $\sigma_f(\mathbf{E}, \mathbf{A}) \neq \emptyset$. Then, from Sect. 3.2.1 for the case of index-3 DAEs, $\boldsymbol{x}$ can be decomposed as

$$\boldsymbol{x} = \begin{pmatrix} \boldsymbol{p}z_{\boldsymbol{p}_0}z_{\boldsymbol{p}_1} & \boldsymbol{p}z_{\boldsymbol{p}_0}z_{\boldsymbol{q}_1} & \boldsymbol{p}z_{\boldsymbol{q}_0} & \boldsymbol{q} \end{pmatrix} \begin{pmatrix} \xi_p \\ \xi_{q,2} \\ \xi_{q,1} \\ \xi_{q,0} \end{pmatrix}, \ \xi_p \in \mathbb{R}^{n_p},\ \xi_{q,2} \in \mathbb{R}^{k_2},\ \xi_{q,1} \in \mathbb{R}^{k_1},\ \xi_{q,0} \in \mathbb{R}^{k_0}.$$

Substituting $\boldsymbol{x}$ into (3.6.1) and simplifying, we obtain

$$\begin{pmatrix}\mathbf{E}_3\mathbf{P}_2\mathbf{P}_1\boldsymbol{p}\boldsymbol{z}_{\boldsymbol{p}_0}\boldsymbol{z}_{\boldsymbol{p}_1} & \mathbf{E}_3\mathbf{P}_2\mathbf{P}_1\boldsymbol{p}\boldsymbol{z}_{\boldsymbol{p}_0}\boldsymbol{z}_{\boldsymbol{q}_1} & -\mathbf{E}_3\mathbf{Q}_0\mathbf{Q}_1\boldsymbol{p}\boldsymbol{z}_{\boldsymbol{q}_0} & 0\end{pmatrix}\begin{pmatrix}\xi_p\\ \xi_{q,2}\\ \xi_{q,1}\\ \xi_{q,0}\end{pmatrix}'$$

$$= \begin{pmatrix}0 & \mathbf{A}_3\boldsymbol{p}\boldsymbol{z}_{\boldsymbol{p}_0}\boldsymbol{z}_{\boldsymbol{p}_1} & -\mathbf{E}_3\mathbf{Q}_2\boldsymbol{p}\boldsymbol{z}_{\boldsymbol{p}_0}\boldsymbol{z}_{\boldsymbol{q}_1} & -\mathbf{E}_3\mathbf{Q}_1\boldsymbol{p}\boldsymbol{z}_{\boldsymbol{q}_0} & -\mathbf{E}_3\boldsymbol{q}_0\end{pmatrix}\begin{pmatrix}\xi_p\\ \xi_{q,2}\\ \xi_{q,1}\\ \xi_{q,0}\end{pmatrix} + \mathbf{B}\mathbf{u} \qquad (3.6.2)$$

Using the same trick as in the previous case, we need to construct another set of bases $\hat{\boldsymbol{p}}_0^T \in \mathbb{R}^{n_0\times n}$, $\hat{\boldsymbol{q}}_0^T \in \mathbb{R}^{k_0\times n}$, $\hat{z}_{\boldsymbol{p}_0}^T \in \mathbb{R}^{k_1,n_0}$, $\hat{z}_{\boldsymbol{q}_0}^T \in \mathbb{R}^{k_0\times n_0}$, $\hat{z}_{\boldsymbol{p}_1}^T \in \mathbb{R}^{n_p\times n_1}$ and $\hat{z}_{\boldsymbol{q}_1}^T \in \mathbb{R}^{k_2\times n_1}$ in order to obtain a one way coupling between the differential and algebraic parts. If we let $\hat{\boldsymbol{p}} \in$ Ker $(\mathbf{E}_3\boldsymbol{q})^T$, $\hat{\boldsymbol{q}} \in$ Ker $(\mathbf{E}_3\boldsymbol{p}_0)^T$, $\hat{z}_{\boldsymbol{p}_0} \in$ Ker $(\hat{\boldsymbol{p}}^T\mathbf{E}_3\boldsymbol{p}\boldsymbol{z}_{\boldsymbol{q}_0})^T\hat{z}_{\boldsymbol{q}_0} \in$ Ker $(\hat{\boldsymbol{p}}^T\mathbf{E}_3\boldsymbol{p}\boldsymbol{z}_{\boldsymbol{p}_0})^T$, $\hat{z}_{\boldsymbol{p}_1} \in$ Ker $(\hat{z}_{\boldsymbol{p}_0}^T\hat{\boldsymbol{p}}^T\mathbf{E}_3\boldsymbol{p}\boldsymbol{z}_{\boldsymbol{p}_0}\boldsymbol{z}_{\boldsymbol{q}_1})^T$ and $\hat{z}_{\boldsymbol{q}_1} \in$ Ker $(\hat{z}_{\boldsymbol{p}_0}^T\hat{\boldsymbol{p}}^T\mathbf{E}_3\boldsymbol{p}\boldsymbol{z}_{\boldsymbol{p}_0}\boldsymbol{z}_{\boldsymbol{p}_1})^T$. Thus left multiplying (3.6.2) by $\begin{pmatrix}\hat{\boldsymbol{p}}\hat{z}_{\boldsymbol{p}_0}\hat{z}_{\boldsymbol{p}_1} & \hat{\boldsymbol{p}}\hat{z}_{\boldsymbol{p}_0}\hat{z}_{\boldsymbol{q}_1} & \hat{\boldsymbol{p}}\hat{z}_{\boldsymbol{q}_0} & \hat{\boldsymbol{q}}\end{pmatrix}^T$ and simplifying leads to an implicit decoupled system of (2.3.1) is given by

$$\begin{aligned}
\mathbf{E}_p\xi_p' &= \mathbf{A}_p\xi_p + \mathbf{B}_p\mathbf{u},\\
\mathbf{E}_{q,2}\xi_{q,2} &= \mathbf{A}_{q,2}\xi_p + \mathbf{B}_{q,2}\mathbf{u},\\
\mathbf{E}_{q,1}\xi_{q,1} &= \mathbf{A}_{q,1}\xi_p + \mathbf{B}_{q,1}\mathbf{u} + \mathbf{E}_{\boldsymbol{q}_{1,2}}\xi_{q,2} + \mathbf{A}_{\boldsymbol{q}_{1,2}}\xi_{q,2}',\\
\mathbf{E}_{q,0}\xi_{q,0} &= \mathbf{A}_{q,0}\xi_p + \mathbf{B}_{q,0}\mathbf{u} + \mathbf{E}_{\boldsymbol{q}_{0,2}}\xi_{q,2} + \mathbf{E}_{\boldsymbol{q}_{0,1}}\xi_{q,1} + \mathbf{A}_{\boldsymbol{q}_{0,2}}\xi_{q,2}' + \mathbf{A}_{\boldsymbol{q}_{0,1}}\xi_{q,1}',\\
\mathbf{y} &= \mathbf{C}_p^T\xi_p + \mathbf{C}_{q,2}^T\xi_{q,2} + \mathbf{C}_{q,1}^T\xi_{q,1} + \mathbf{C}_{q,0}^T\xi_{q,0},
\end{aligned} \qquad (3.6.3)$$

where
$\mathbf{E}_p = \hat{z}_{\boldsymbol{p}_1}^T\hat{z}_{\boldsymbol{p}_0}^T\hat{\boldsymbol{p}}^T\mathbf{E}_3\mathbf{P}_2\mathbf{P}_1\boldsymbol{p}\boldsymbol{z}_{\boldsymbol{p}_0}\boldsymbol{z}_{\boldsymbol{p}_1} \in \mathbb{R}^{n_p\times n_p}$, $\mathbf{A}_p = \hat{z}_{\boldsymbol{p}_1}^T\hat{z}_{\boldsymbol{p}_0}^T\hat{\boldsymbol{p}}^T\mathbf{A}_3\boldsymbol{p}\boldsymbol{z}_{\boldsymbol{p}_0}\boldsymbol{z}_{\boldsymbol{p}_1} \in \mathbb{R}^{n_p\times n_p}$, $\mathbf{B}_p = \hat{z}_{\boldsymbol{p}_1}^T\hat{z}_{\boldsymbol{p}_0}^T\hat{\boldsymbol{p}}^T\mathbf{B} \in \mathbb{R}^{n_p\times m}$,
$\mathbf{E}_{q,2} = \hat{z}_{\boldsymbol{q}_1}^T\hat{z}_{\boldsymbol{p}_0}^T\hat{\boldsymbol{p}}^T\mathbf{E}_3\mathbf{Q}_2\boldsymbol{p}\boldsymbol{z}_{\boldsymbol{p}_0}\boldsymbol{z}_{\boldsymbol{q}_1} \in \mathbb{R}^{k_2\times k_2}$, $\mathbf{A}_{q,2} = \hat{z}_{\boldsymbol{q}_1}^T\hat{z}_{\boldsymbol{p}_0}^T\hat{\boldsymbol{p}}^T\mathbf{A}_3\boldsymbol{p}\boldsymbol{z}_{\boldsymbol{p}_0}\boldsymbol{z}_{\boldsymbol{p}_1} \in \mathbb{R}^{k_2\times n_p}$, $\mathbf{B}_{q,2} = \hat{z}_{\boldsymbol{q}_1}^T\hat{z}_{\boldsymbol{p}_0}^T\hat{\boldsymbol{p}}^T\mathbf{B} \in \mathbb{R}^{k_2\times m}$,
$\mathbf{E}_{q,1} = \hat{z}_{\boldsymbol{q}_0}^T\hat{\boldsymbol{p}}^T\mathbf{E}_3\mathbf{Q}_1\boldsymbol{p}\boldsymbol{z}_{\boldsymbol{q}_0} \in \mathbb{R}^{k_1\times k_1}$, $\mathbf{A}_{q,1} = \hat{z}_{\boldsymbol{q}_0}^T\hat{\boldsymbol{p}}^T\mathbf{A}_3\boldsymbol{p}\boldsymbol{z}_{\boldsymbol{p}_0}\boldsymbol{z}_{\boldsymbol{p}_1} \in \mathbb{R}^{k_1\times n_p}$, $\mathbf{B}_{q,1} = \hat{z}_{\boldsymbol{q}_0}^T\hat{\boldsymbol{p}}^T\mathbf{B} \in \mathbb{R}^{k_1\times m}$,
$\mathbf{E}_{\boldsymbol{q}_{1,2}} = -\hat{z}_{\boldsymbol{q}_0}^T\hat{\boldsymbol{p}}^T\mathbf{E}_3\mathbf{Q}_2\boldsymbol{p}\boldsymbol{z}_{\boldsymbol{p}_0}\boldsymbol{z}_{\boldsymbol{q}_1} \in \mathbb{R}^{k_1\times k_2}$, $\mathbf{A}_{\boldsymbol{q}_{1,2}} = -\hat{z}_{\boldsymbol{q}_0}^T\hat{\boldsymbol{p}}^T\mathbf{E}_3\mathbf{P}_2\mathbf{P}_1\boldsymbol{p}\boldsymbol{z}_{\boldsymbol{p}_0}\boldsymbol{z}_{\boldsymbol{q}_1} \in \mathbb{R}^{k_1\times k_2}$, $\mathbf{E}_{q,0} = \hat{\boldsymbol{q}}^T\mathbf{E}_3\boldsymbol{q} \in \mathbb{R}^{k_0\times k_0}$,
$\mathbf{A}_{q,0} = \hat{\boldsymbol{q}}^T\mathbf{A}_3\boldsymbol{p}\boldsymbol{z}_{\boldsymbol{p}_0}\boldsymbol{z}_{\boldsymbol{p}_1} \in \mathbb{R}^{k_0\times n_p}$, $\mathbf{E}_{\boldsymbol{q}_{0,2}} = -\hat{\boldsymbol{q}}^T\mathbf{E}_3\mathbf{Q}_2\boldsymbol{p}\boldsymbol{z}_{\boldsymbol{p}_0}\boldsymbol{z}_{\boldsymbol{q}_1} \in \mathbb{R}^{k_0\times k_2}$, $\mathbf{B}_{q,0} = \hat{\boldsymbol{q}}^T\mathbf{B} \in \mathbb{R}^{k_0\times m}$,
$\mathbf{E}_{\boldsymbol{q}_{0,1}} = -\hat{\boldsymbol{q}}^T\mathbf{E}_3\mathbf{Q}_1\boldsymbol{p}\boldsymbol{z}_{\boldsymbol{q}_0} \in \mathbb{R}^{k_0\times k_1}$, $\mathbf{A}_{\boldsymbol{q}_{0,2}} = -\hat{\boldsymbol{q}}^T\mathbf{E}_3\mathbf{P}_2\mathbf{P}_1\boldsymbol{p}\boldsymbol{z}_{\boldsymbol{p}_0}\boldsymbol{z}_{\boldsymbol{q}_1} \in \mathbb{R}^{k_0\times k_2}$, $\mathbf{C}_{q,1}^T = \mathbf{C}^T\boldsymbol{p}_0\boldsymbol{z}_{\boldsymbol{q}_0} \in \mathbb{R}^{k_1\times \ell}$,
$\mathbf{C}_{q,2}^T = \mathbf{C}^T\boldsymbol{p}_0\boldsymbol{z}_{\boldsymbol{p}_0}\boldsymbol{z}_{\boldsymbol{q}_1} \in \mathbb{R}^{k_2\times \ell}$, $\mathbf{A}_{\boldsymbol{q}_{0,1}} = \hat{\boldsymbol{q}}^T\mathbf{E}_3\mathbf{Q}_0\mathbf{Q}_1\boldsymbol{p}\boldsymbol{z}_{\boldsymbol{q}_0} \in \mathbb{R}^{k_0\times k_1}$, $\mathbf{C}_{q,0}^T = \mathbf{C}^T\boldsymbol{q}_0 \in \mathbb{R}^{k_0\times \ell}$,
and $\mathbf{C}_p^T = \mathbf{C}^T\boldsymbol{p}_0\boldsymbol{z}_{\boldsymbol{p}_0}\boldsymbol{z}_{\boldsymbol{p}_1} \in \mathbb{R}^{n_p\times \ell}$.

3.6.2 Index-3 DAEs with Only Infinite Spectrum

Here, we assume that the matrix pencil $(\mathbf{E}, \mathbf{A})$ of (2.3.1) has no finite eigenvalues, i.e., $\sigma_f(\mathbf{E}, \mathbf{A}) = \emptyset$. This implies that $\mathbf{P}_0\mathbf{P}_1\mathbf{P}_2 = 0$, thus (3.6.1) simplifies to,

$$\mathbf{E}_3\Big[\mathbf{P}_2\mathbf{P}_1\mathbf{P}_0\boldsymbol{x}' + \mathbf{Q}_2\boldsymbol{x} + \mathbf{Q}_1\boldsymbol{x} + \mathbf{Q}_0\boldsymbol{x}\Big] = \mathbf{Bu}, \tag{3.6.4}$$

Also from Sect. 3.2.2 for the case of index-3 DAEs with only infinite spectrum, $\boldsymbol{x}$ can be decomposed as $\boldsymbol{x} = \begin{pmatrix} \boldsymbol{pz_{p_0}} & \boldsymbol{pz_{q_0}} & \boldsymbol{q} \end{pmatrix} \begin{pmatrix} \xi_{q,2} \\ \xi_{q,1} \\ \xi_{q,0} \end{pmatrix}$, $\xi_{q,2} \in \mathbb{R}^{k_2}$, $\xi_{q,1} \in \mathbb{R}^{k_1}$, $\xi_{q,0} \in \mathbb{R}^{k_0}$. Then substituting $\boldsymbol{x}$ into (3.6.4) and simplifying we obtain

$$\begin{pmatrix} \mathbf{E}_3\mathbf{P}_2\boldsymbol{pz_{p_0}} & -\mathbf{E}_3\mathbf{Q}_0\mathbf{Q}_1\boldsymbol{pz_{q_0}} & 0 \end{pmatrix} \begin{pmatrix} \xi_{q,2} \\ \xi_{q,1} \\ \xi_{q,0} \end{pmatrix}' = \begin{pmatrix} -\mathbf{E}_3\mathbf{Q}_2\boldsymbol{pz_{p_0}} & -\mathbf{E}_3\mathbf{Q}_1\boldsymbol{pz_{q_0}} & -\mathbf{E}_3\boldsymbol{q}_0 \end{pmatrix} \begin{pmatrix} \xi_{q,2} \\ \xi_{q,1} \\ \xi_{q,0} \end{pmatrix} + \mathbf{Bu}. \tag{3.6.5}$$

Using the same trick as in the previous case, we need to construct another set of bases $\hat{\boldsymbol{p}}^T \in \mathbb{R}^{n_0 \times n}$, $\hat{\boldsymbol{q}}^T \in \mathbb{R}^{k_0 \times n}$, $\hat{\boldsymbol{z}}_{\boldsymbol{p}_0}^T \in \mathbb{R}^{n_1 \times n_0}$, $\hat{\boldsymbol{z}}_{\boldsymbol{q}_0}^T \in \mathbb{R}^{k_1 \times n_0}$. If we let $\hat{\boldsymbol{p}} \in \text{Ker}\,(\mathbf{E}_3\boldsymbol{q})^T$, $\hat{\boldsymbol{q}} \in \text{Ker}\,(\mathbf{E}_3\boldsymbol{p}_0)^T$, $\text{Span}(\hat{z}_{\boldsymbol{p}_0}) = \text{Ker}\,(\hat{\boldsymbol{p}}^T\mathbf{E}_3\mathbf{Q}_1\boldsymbol{pz_{q_0}})^T$ and $\text{Span}(\hat{z}_{\boldsymbol{q}_0}) = \text{Ker}\,(\hat{\boldsymbol{p}}^T\mathbf{E}_3\mathbf{Q}_2\boldsymbol{pz_{p_0}})^T$. Thus left multiplying (3.6.5) by $\begin{pmatrix} \hat{\boldsymbol{p}}\hat{z}_{\boldsymbol{p}_0} & \hat{\boldsymbol{p}}\hat{z}_{\boldsymbol{q}_0} & \hat{\boldsymbol{q}} \end{pmatrix}^T$ and simplifying, we obtain the implicit decoupled system of (2.3.1) is given by

$$\begin{aligned} \mathbf{E}_{q,2}\xi_{q,2} &= \mathbf{B}_{q,2}\mathbf{u}, \\ \mathbf{E}_{q,1}\xi_{q,1} &= \mathbf{B}_{q,1}\mathbf{u} + \mathbf{A}_{\boldsymbol{q}_{1,2}}\xi'_{q,2}, \\ \mathbf{E}_{q,0}\xi_{q,0} &= \mathbf{B}_{q,0}\mathbf{u} + \mathbf{E}_{\boldsymbol{q}_{0,2}}\xi_{q,2} + \mathbf{E}_{\boldsymbol{q}_{0,1}}\xi_{q,1} + \mathbf{A}_{\boldsymbol{q}_{0,2}}\xi'_{q,2} + \mathbf{A}_{\boldsymbol{q}_{0,1}}\xi'_{q,1}, \\ \mathbf{y} &= \mathbf{C}_{q,2}^T\xi_{q,2} + \mathbf{C}_{q,1}^T\xi_{q,1} + \mathbf{C}_{q,0}^T\xi_{q,0}, \end{aligned} \tag{3.6.6}$$

where $\mathbf{E}_{q,2} = \hat{z}_{\boldsymbol{p}_0}^T\hat{\boldsymbol{p}}^T\mathbf{E}_3\mathbf{Q}_2\boldsymbol{pz_{p_0}} \in \mathbb{R}^{k_2 \times k_2}$, $\mathbf{B}_{q,2} = \hat{z}_{\boldsymbol{p}_0}^T\hat{\boldsymbol{p}}^T\mathbf{B} \in \mathbb{R}^{k_2 \times m}$, $\mathbf{E}_{q,1} = \hat{z}_{\boldsymbol{q}_0}^T\hat{\boldsymbol{p}}^T\mathbf{E}_3\mathbf{Q}_1\boldsymbol{pz_{q_0}} \in \mathbb{R}^{k_1 \times k_1}$, $\mathbf{B}_{q,1} = \hat{z}_{\boldsymbol{q}_0}^T\hat{\boldsymbol{p}}^T\mathbf{B} \in \mathbb{R}^{k_1 \times m}$, $\mathbf{A}_{\boldsymbol{q}_{1,2}} = -\hat{z}_{\boldsymbol{q}_0}^T\hat{\boldsymbol{p}}^T\mathbf{E}_3\mathbf{P}_2\boldsymbol{pz_{p_0}} \in \mathbb{R}^{k_1 \times k_2}$, $\mathbf{E}_{q,0} = \hat{\boldsymbol{q}}^T\mathbf{E}_3\boldsymbol{q}_0 \in \mathbb{R}^{k_0,k_0}$, $\mathbf{B}_{q,0} = \hat{\boldsymbol{q}}^T\mathbf{B} \in \mathbb{R}^{k_0 \times m}$, $\mathbf{A}_{\boldsymbol{q}_{0,2}} = -\hat{\boldsymbol{q}}^T\mathbf{E}_3\mathbf{P}_2\boldsymbol{pz_{p_0}} \in \mathbb{R}^{k_0 \times k_2}$, and $\mathbf{A}_{\boldsymbol{q}_{0,1}} = \hat{\boldsymbol{q}}^T\mathbf{E}_3\mathbf{Q}_0\mathbf{Q}_1\boldsymbol{pz_{q_0}} \in \mathbb{R}^{k_0 \times k_1}$, $\mathbf{E}_{\boldsymbol{q}_{0,1}} = -\hat{\boldsymbol{q}}^T\mathbf{E}_3\mathbf{Q}_1\boldsymbol{pz_{q_0}} \in \mathbb{R}^{k_0 \times k_1}$, $\mathbf{E}_{\boldsymbol{q}_{0,2}} = -\hat{\boldsymbol{q}}^T\mathbf{E}_3\mathbf{Q}_2\boldsymbol{pz_{p_0}} \in \mathbb{R}^{k_0 \times k_2}$, $\mathbf{C}_{q,2}^T = \mathbf{C}^T\boldsymbol{p}_0\boldsymbol{z_{p_0}} \in \mathbb{R}^{k_2 \times \ell}$, $\mathbf{C}_{q,1}^T = \mathbf{C}^T\boldsymbol{p}_0\boldsymbol{z_{q_0}} \in \mathbb{R}^{k_1 \times \ell}$, $\mathbf{C}_{q,0}^T = \mathbf{C}^T\boldsymbol{q}_0 \in \mathbb{R}^{k_0 \times \ell}$.

In the next examples, we illustrate how to derive implicit decoupled systems. For comparison, we use the same DAE matrices as Examples 3.2.1 and 3.2.2.

Example 3.6.1 Here, we use matrix, projector and right basis chain from Example 3.2.1. This is done as follows. Since this is an index-1 DAE its decoupled system will take the form (3.4.6). Using matrices (3.2.25) and right bases (3.2.30) for projector $\mathbf{Q}_0$ and $\mathbf{P}_0$, we were to able to construct left bases given by

$$\hat{p}_0 = \begin{pmatrix} 0 & 0 \\ 0 & 1/G \\ 1 & 0 \\ 0 & 1 \\ 0 & 1 \end{pmatrix} \in \operatorname{Ker} q_0^T \mathbf{A}^T \quad \text{and} \quad \hat{q}_0 = \begin{pmatrix} 1 & 0 & 0 \\ 0 & 1 & 0 \\ 0 & 0 & 0 \\ 0 & 0 & 0 \\ 0 & 0 & 1 \end{pmatrix} \in \operatorname{Ker} \mathbf{E}^T. \tag{3.6.7}$$

Then, substituting the above left bases, right bases (3.2.30) and system matrices (3.2.25) into (3.4.6), we obtain an equivalent implicit decoupled system given by

$$\begin{pmatrix} C & 0 \\ 0 & L \end{pmatrix} \xi_p' = \begin{pmatrix} 0 & 1 \\ -1 & -1/G \end{pmatrix} \xi_p + \begin{pmatrix} 0 \\ -1 \end{pmatrix} \boldsymbol{u}, \tag{3.6.8a}$$

$$\begin{pmatrix} G & -G & -1 \\ -G & G & 0 \\ 1 & 0 & 0 \end{pmatrix} \xi_{q,0} = \begin{pmatrix} 0 & 0 \\ 0 & 0 \\ 0 & 0 \end{pmatrix} \xi_p + \begin{pmatrix} 0 \\ 0 \\ -1 \end{pmatrix} \boldsymbol{u}, \tag{3.6.8b}$$

$$\mathbf{y} = \begin{pmatrix} 0 & -1 \end{pmatrix} \xi_p + \begin{pmatrix} 0 & 0 & -1 \end{pmatrix} \xi_{q,0}. \tag{3.6.8c}$$

We can observe that the above decoupled system is equivalent to (3.2.32) and there solution coincides to $\mathbf{y} = -\xi_{p_2}$.

Example 3.6.2 Here, we use matrix,projector and right basis chain from Example 3.2.2. This is done as follows. Since this is an index-2 DAE with a finite spectrum its decoupled system will take the form (3.5.4). Using using matrices (3.2.42) and right bases (3.2.43), we obtain the first left basis given by

$$\hat{p}_0 = \begin{pmatrix} 0 & 0 \\ 1/G & 0 \\ 1 & 0 \\ 0 & 1 \end{pmatrix} \in \operatorname{Ker} q_0^T \mathbf{E}_2^T \quad \text{and} \quad \hat{q}_0 = \begin{pmatrix} \frac{G}{C+G} & -\frac{1}{C+G} \\ 1 & 0 \\ 0 & 0 \\ 0 & 1 \end{pmatrix} \in \operatorname{Ker} (p_0^T \mathbf{E}_2^T. \tag{3.6.9}$$

Using the above left bases and matrices (3.2.42)–(3.2.44), we obtain the second set of left bases given by

$$\hat{\mathbf{z}}_{p_0} = \begin{pmatrix} 1 \\ 1 \end{pmatrix} \in \operatorname{Ker} (\hat{p}_0^T \mathbf{E}_2^T p_0 \mathbf{z}_{q_0})^T \quad \text{and} \quad \hat{\mathbf{z}}_{q_0} = \begin{pmatrix} 0 \\ 1 \end{pmatrix} \in \operatorname{Ker} (\hat{p}_0^T \mathbf{E}_2^T p_0 \mathbf{z}_{p_0})^T. \tag{3.6.10}$$

Substituting the above left bases and matrices (3.2.42) and (3.2.44) into (3.5.4), we obtain an equivalent implicit decoupled system of (3.2.33) given by

$$L\xi_p' = -1/G\xi_p - \boldsymbol{u}, \tag{3.6.11a}$$

$$\xi_{q,1} = 0\xi_p - \mathbf{u}, \tag{3.6.11b}$$

$$\frac{1}{C+G} \begin{pmatrix} -CG & G \\ -G & -1 \end{pmatrix} \xi_{q,0} = \begin{pmatrix} 0 \\ 0 \end{pmatrix} \xi_p + \begin{pmatrix} 0 \\ -1 \end{pmatrix} \mathbf{u} + \frac{1}{C+G} \begin{pmatrix} -CG \\ C \end{pmatrix} \left[\xi_{q,1}' - \xi_{q,1} \right], \tag{3.6.11c}$$

$$\mathbf{y} = -\xi_p + 0\xi_{q,1} + \begin{pmatrix} 0 & -1 \end{pmatrix} \xi_{q,0} \tag{3.6.11d}$$

If we compare (3.2.45) and (3.6.11), we can observe that the two systems are equivalent and their output solution coincide to $\mathbf{y} = -\xi_p + C\mathbf{u}'$.

Example 3.6.3 In this example, we use system matrices from Example 3.2.3. This DAE has matrix pencil whose spectrum as no finite eigenvalues and it is of index-2. Thus, its implicit decoupled system will take the form (3.5.7). Using Eqs. (3.2.47)–(3.2.48), we constructed the left bases given by $\hat{\boldsymbol{p}}_0 = \begin{pmatrix} 1 \\ 1 \\ 0 \end{pmatrix} \in \operatorname{Ker} \boldsymbol{q}_0^T \mathbf{E}_2^T$ and $\hat{\boldsymbol{q}}_0 = \begin{pmatrix} 1 & 0 \\ 0 & 0 \\ 0 & 1 \end{pmatrix} \in \operatorname{Ker} \boldsymbol{p}_0^T \mathbf{Q}_1^T \mathbf{E}_2^T$. Substituting the above bases and Eqs. (3.2.47)–(3.2.48) into (3.5.7), we obtain an implicit decoupled system of (3.2.46) given by

$$
\begin{aligned}
\xi_{q,1} &= \mathbf{u}, \\
\begin{pmatrix} G & -G \\ 0 & -1 \end{pmatrix} \xi_{q,0} &= \begin{pmatrix} 1 \\ 0 \end{pmatrix} \mathbf{u} + \begin{pmatrix} 0 \\ -L \end{pmatrix} \xi_{q,1}', \\
\mathbf{y} &= 0\xi_{q,1} + \begin{pmatrix} 1 & 0 \end{pmatrix} \xi_{q,0}.
\end{aligned}
\tag{3.6.12}
$$

We can observe that the above decoupled system coincides with (3.2.49) and their output solution coincides to $\mathbf{y} = G^{-1}\mathbf{u} + L\mathbf{u}'$.

Example 3.6.4 In this example, we use matrices from Example 3.2.4. This DAE is of index-3 and its matrix has at least one finite eigenvalue. Thus, its implicit decoupled system will take the form (3.6.3). Thus following the procedure in Sect. 3.6.1, we can construct the left bases as follows. Using Eqs. (3.2.52) and (3.2.53), we can obtain the first left bases given by

$$
\hat{\boldsymbol{p}}_0 = \begin{pmatrix}
1 & 0 & 0 & 0 & 0 & 0 & 0 \\
0 & 1 & 0 & 0 & 0 & 0 & 0 \\
0 & 0 & 1 & 0 & 0 & 0 & 0 \\
0 & 0 & 0 & 1 & 0 & 0 & 0 \\
0 & 0 & 0 & 0 & 1 & 0 & 0 \\
0 & 0 & 0 & 0 & 0 & 0 & 1
\end{pmatrix}^T \in \operatorname{Ker} (\mathbf{E}_3 \boldsymbol{q})^T,
$$

$$
\hat{\boldsymbol{q}} = \begin{pmatrix} 0 & 0 & -\frac{4L^2}{3L+g} & 0 & 0 & -\frac{2L^2}{m_2(3L+g)} & 1 \end{pmatrix}^T \in \operatorname{Ker} (\mathbf{E}_3 \boldsymbol{p}_0)^T.
$$

Next, we use the above bases and (3.2.54) to obtain the second set of left bases given by

$$\hat{z}_{p_0} = \begin{pmatrix} 1\,0\,0\,0\,0\,0 \\ 0\,1\,0\,0\,0\,0 \\ 0\,0\,0\,1\,0\,0 \\ 0\,0\,0\,0\,1\,0 \\ 0\,0\,0\,0\,0\,1 \end{pmatrix}^T \in \text{Ker}\,(\hat{\boldsymbol{p}}^T \mathbf{E}_3 \boldsymbol{p} z_{q_0})^T, \hat{z}_{q_0} = \begin{pmatrix} 0 \\ 0 \\ -L \\ 0 \\ 0 \\ 1 \end{pmatrix} \in \text{Ker}\,(\hat{\boldsymbol{p}}^T \mathbf{E}_3 \boldsymbol{p} z_{p_0})^T,$$

Finally, we use the above bases and (3.2.54), we obtain the third set of left bases given by

$$\hat{z}_{p_1} = \begin{pmatrix} 1\,0\,0\,0\,0 \\ 0\,1\,0\,0\,0 \\ 0\,0\,1\,0\,0 \\ 0\,0\,0\,1\,0 \end{pmatrix}^T \in \text{Ker}\,(\hat{z}_{p_0}^T \hat{\boldsymbol{p}}^T \mathbf{E}_3 \boldsymbol{p} z_{p_0} z_{q_1})^T,$$

$$\hat{z}_{q_1} = \begin{pmatrix} 0 \\ 0 \\ 0 \\ 0 \\ 1 \end{pmatrix} \in \text{Ker}\,(\hat{z}_{p_0}^T \hat{\boldsymbol{p}}^T \mathbf{E}_3 \boldsymbol{p} z_{p_0} z_{p_1})^T.$$

Substituting the above column matrices and those from Eqs. (3.2.50)–(3.2.55) into (3.6.3), we obtain an equivalent implicit decoupled system of (3.2.50) given by

$$\begin{aligned}
\begin{pmatrix} 1 & 0 & 0 & 0 \\ 0 & 1 & 0 & 0 \\ 0 & 0 & m_1 & 0 \\ 0 & 0 & 0 & m_2 \end{pmatrix} \xi_p' &= \begin{pmatrix} 0 & 0 & 1 & 0 \\ 0 & 0 & 0 & 1 \\ -\frac{gm_2}{L} & \frac{gm_2}{L} & 0 & 0 \\ \frac{gm_2}{L} & -\frac{gm_2}{L} & 0 & 0 \end{pmatrix} \xi_p + \begin{pmatrix} 0 \\ 0 \\ 1 \\ 0 \end{pmatrix} \mathbf{u}, \\
-2L\xi_{q,2} &= 0\xi_p + 0\mathbf{u}, \\
-\frac{2L^2}{m_2}\xi_{q,1} &= \begin{pmatrix} 0 & 0 & 0 & 0 \end{pmatrix} \xi_p + 0\mathbf{u} + L\xi_{q,2} - L\xi_{q,2}', \\
\frac{4L^3}{m_2(3L+g)}\xi_{q,0} &= \begin{pmatrix} 0 & 0 & 0 & 0 \end{pmatrix} \xi_p + 0\mathbf{u} + \frac{2L^2}{(3L+g)}\xi_{q,2} + -\frac{2L^2}{(3L+g)}\xi_{q,2}' + \frac{4L^3}{m_2(3L+g)}\left[\xi_{q,1} - \xi_{q,1}'\right], \\
\mathbf{y} &= \begin{pmatrix} 0\,1\,0\,0 \\ 0\,0\,0\,0 \end{pmatrix} \xi_p + \begin{pmatrix} 0 \\ 1 \end{pmatrix} \xi_{q,2} + \begin{pmatrix} 0 \\ 0 \end{pmatrix} \xi_{q,1} + \begin{pmatrix} 0 \\ 0 \end{pmatrix} \xi_{q,0}.
\end{aligned}$$

We can observe that the above decoupled system can be simplified to an ODE system given by

$$\begin{aligned}
\begin{pmatrix} 1 & 0 & 0 & 0 \\ 0 & 1 & 0 & 0 \\ 0 & 0 & m_1 & 0 \\ 0 & 0 & 0 & m_2 \end{pmatrix} \xi_p' &= \begin{pmatrix} 0 & 0 & 1 & 0 \\ 0 & 0 & 0 & 1 \\ -\frac{gm_2}{L} & \frac{gm_2}{L} & 0 & 0 \\ \frac{gm_2}{L} & -\frac{gm_2}{L} & 0 & 0 \end{pmatrix} \xi_p + \begin{pmatrix} 0 \\ 0 \\ 1 \\ 0 \end{pmatrix} \mathbf{u}, \\
\mathbf{y} &= \begin{pmatrix} 0\,1\,0\,0 \\ 0\,0\,0\,0 \end{pmatrix} \xi_p,
\end{aligned} \tag{3.6.13}$$

We can observe (3.6.13) is an implicit version of (3.2.56) and their solutions coincide.

Example 3.6.5 In this example, we use matrices from Example 3.2.5. We can recall that the spectrum of the matrix pencil of this DAE is index-3 and has only infinite spectrum. Thus, its implicit decoupled system is in the form (3.6.6). Following procedure in Sect. 3.6.2 and using matrices from 3.2.5, we can construct bases given by

$$\hat{\boldsymbol{p}}_0 = \begin{pmatrix} 0 & 0 \\ 1 & 0 \\ 0 & 1 \end{pmatrix} \in \text{Ker}\,(\mathbf{E}_3\boldsymbol{q})^T,\ \hat{\boldsymbol{q}}_0 = \begin{pmatrix} \frac{1}{3} \\ \frac{2}{3} \\ 1 \end{pmatrix} \in \text{Ker}\,(\mathbf{E}_3\boldsymbol{p}_0)^T,$$

$$\hat{z}_{p_0} = \begin{pmatrix} 0 \\ 1 \end{pmatrix} \in \text{Ker}\,(\hat{\boldsymbol{p}}^T\mathbf{E}_3\mathbf{Q}_1\boldsymbol{p}z_{q_0})^T,\ \hat{z}^T_{q_0} = \begin{pmatrix} \frac{1}{\sqrt{2}} \\ \frac{1}{\sqrt{2}} \end{pmatrix} \in \text{Ker}\,(\hat{\boldsymbol{p}}^T\mathbf{E}_3\mathbf{Q}_2\boldsymbol{p}z_{p_0})^T.$$

Substituting these column matrices and those from Example 3.2.5 into (3.6.6), we obtain an implicit decoupled system of (3.2.57) given by

$$\begin{aligned} -\xi_{q,2} &= 0\mathbf{u}, \\ -\frac{1}{\sqrt{2}}\xi_{q,1} &= \frac{1}{10\sqrt{2}}\mathbf{u} - \frac{1}{\sqrt{2}}\xi'_{q,2}, \\ -\frac{1}{3}\xi_{q,0} &= \frac{17}{5}\mathbf{u} + \frac{1}{3}\xi_{q,2} + \frac{1}{3}\xi_{q,1} - \frac{1}{3}\xi'_{q,2} - \frac{1}{3}\xi'_{q,1}, \\ \mathbf{y} &= -29\xi_{q,2} + 29.96\xi_{q,1} + 0.04\xi_{q,0}. \end{aligned}$$

After solving the above system leads to an output solution $\mathbf{y} = -3.4\boldsymbol{u} - 0.004\boldsymbol{u}'$ which coincides with that obtained in Example 3.2.5.

3.7 Comparison of Decoupling Methods

In this section, we compare the decoupling methods for DAEs discussed in Sects. 3.2 and 3.3. Without loss of generality, if the DAE (2.3.1) is of index-μ and its matrix pencil has at least one finite eigenvalue. Then, following the decoupling method in Sect. 3.3 the equivalent implicit decoupled system of (2.3.1) is given by

$$\mathbf{E}_p\xi'_p = \mathbf{A}_p\xi_p + \mathbf{B}_p\mathbf{u} \tag{3.7.1a}$$

$$-\mathcal{L}\xi'_q = \mathbf{A}_q\xi_p - \mathcal{L}_q\xi_q + \mathbf{B}_q\mathbf{u}, \tag{3.7.1b}$$

$$\mathbf{y} = \mathbf{C}_p^T\xi_p + \mathbf{C}_q^T\xi_q, \tag{3.7.1c}$$

where $\mathcal{L}$ is a nilpotent matrix of index μ. $\mathcal{L}_q$ is a non-singular lower triangular matrix with block diagonal matrices for $\mu > 1$. $\xi_p \in \mathbb{R}^{n_p}$, $\xi_q \in \mathbb{R}^{n_q}$, $\mathbf{A}_q \in \mathbb{R}^{n_q \times n_p}$, $\mathbf{B}_q \in \mathbb{R}^{n_q \times m}$ and $\mathbf{C}_q \in \mathbb{R}^{n_q \times \ell}$, $\mathbf{C}_p \in \mathbb{R}^{n_p \times \ell}$. If spectrum of the matrix pencil of (2.3.1) has only infinite eigenvalues then the equivalent implicit decoupled system of (2.3.1) is purely algebraic given by

$$-\mathcal{L}\xi_q' = -\mathcal{L}_q\xi_q + \mathbf{B}_q\mathbf{u}, \tag{3.7.2a}$$
$$\mathbf{y} = \mathbf{C}_q^T\,\xi_q, \tag{3.7.2b}$$

For comparison with the DAE (2.3.1), we can rewrite the implicit decoupled systems either (3.7.1) or (3.7.2) in the descriptor form given by

$$\tilde{\mathbf{E}}\xi' = \tilde{\mathbf{A}}\xi + \tilde{\mathbf{B}}\mathbf{u}, \tag{3.7.3a}$$
$$\mathbf{y} = \tilde{\mathbf{C}}^T\xi, \tag{3.7.3b}$$

where $\tilde{\mathbf{E}} = \begin{pmatrix} \mathbf{E}_p & 0 \\ 0 & -\mathcal{L} \end{pmatrix} \in \mathbb{R}^{n\times n}$, $\tilde{\mathbf{A}} = \begin{pmatrix} \mathbf{A}_p & 0 \\ \mathbf{A}_q & -\mathcal{L}_q \end{pmatrix} \in \mathbb{R}^{n\times n}$, $\tilde{\mathbf{B}} = \begin{pmatrix} \mathbf{B}_p \\ \mathbf{B}_q \end{pmatrix} \in \mathbb{R}^{n\times m}$, $\tilde{\mathbf{C}} = \begin{pmatrix} \mathbf{C}_p \\ \mathbf{C}_q \end{pmatrix} \in \mathbb{R}^{n\times \ell}$ if the spectrum of the matrix pencil $(\mathbf{E}, \mathbf{A})$ has at least one finite eigenvalue and $\tilde{\mathbf{E}} = -\mathcal{L} \in \mathbb{R}^{n\times n}$, $\tilde{\mathbf{A}} = -\mathcal{L}_q \in \mathbb{R}^{n\times n}$, $\tilde{\mathbf{B}} = \mathbf{B}_q \in \mathbb{R}^{n\times m}$, $\tilde{\mathbf{C}} = \mathbf{C}_q \in \mathbb{R}^{n\times \ell}$, if the spectrum of the matrix pencil $(\mathbf{E}, \mathbf{A})$ has no finite eigenvalues. We can observe that this form also reveals the interconnection structure of the DAE (2.3.1). Moreover it can also be proved that systems (2.3.1) and (3.7.3) are equivalent. This implies that also their matrix pencils $(\mathbf{E}, \mathbf{A})$ and $(\tilde{\mathbf{E}}, \tilde{\mathbf{A}})$ are equivalent, thus they must have the same spectrum.

If we consider DAEs whose matrix pencil $(\mathbf{E}, \mathbf{A})$ has at least one finite eigenvalue, we can easily show that $\det(\lambda\tilde{\mathbf{E}} - \tilde{\mathbf{A}}) = \det(\lambda\mathbf{E}_p - \mathbf{A}_p)$, since $\det(\mathcal{L}_q - \lambda\mathcal{L}) = (1)^{n_q}$. This identity shows that the finite eigenvalues of the matrix pencil $(\mathbf{E}, \mathbf{A})$ coincide with the (possibly complex) eigenvalues of the matrix $\mathbf{E}_p^{-1}\mathbf{A}_p$ of the differential part, which are exactly n_p, counting their multiplicity. This implies that the differential part of the implicit decoupled system also inherits the stability properties of the DAE (2.3.1). If we compare the descriptor forms (3.2.24) and (3.7.3), they coincide if $\mathbf{E}_p = \mathrm{I}$ and $\mathcal{L}_q = \mathrm{I}$ their systems are equivalent. Hence both the implicit and explicit decoupling procedure have the same mathematical properties. However, if we choose canonical projectors in advance, we can always have a completely decoupling for decoupled system (3.2.24) that is $\mathbf{A}_q = 0$ while for the case of (3.7.3) it only happens if singular matrix $\mathbf{E}$ is symmetric. Another, main difference between these two procedures is the computational cost involved in deriving the respective decoupled systems. The explicit decoupling is the most expensive since its decoupling procedure involves inversion of matrix $\mathbf{E}_\mu$ which can be computationally very expensive.

3.8 Decoupled System for DAEs with Special Structures

In this section, we discuss how to decoupled systems from DAEs with special structures that arises from real-life applications such as computational electromagnetism, electrical networks, computational fluid dynamics (CFD) etc.

Example 3.8.1 Consider a semi-explicit index-1 DAE of the for (2.3.1) with system matrices

$$\mathbf{E} = \begin{pmatrix} \mathbf{E}_{11} & \mathbf{E}_{12} \\ \mathbf{0} & \mathbf{0} \end{pmatrix}, \ \mathbf{A} = \begin{pmatrix} \mathbf{A}_{11} & \mathbf{A}_{12} \\ \mathbf{A}_{21} & \mathbf{A}_{22} \end{pmatrix}, \ \mathbf{B} = \begin{pmatrix} \mathbf{B}_1 \\ \mathbf{B}_2 \end{pmatrix}, \ \mathbf{C} = \begin{pmatrix} \mathbf{C}_1 \\ \mathbf{C}_2 \end{pmatrix}. \tag{3.8.1}$$

We assume $\mathbf{E}_{11} \in \mathbb{R}^{n_1,n_1}$ and $\mathbf{A}_{21}\mathbf{E}_{11}^{-1}\mathbf{E}_{12} - \mathbf{A}_{22} \in \mathbb{R}^{n_2 \times n_2}$ are non-singular blocks due to index-1 property and $n = n_1 + n_2$ is the dimension of the DAE. We can choose projector $\mathbf{Q}_0 = \begin{pmatrix} \mathbf{0} & -\mathbf{Q}_{12} \\ \mathbf{0} & \mathrm{I} \end{pmatrix}$ and its complementary projector $\mathbf{P}_0 = \mathrm{I} - \mathbf{Q} = \begin{pmatrix} \mathrm{I} & \mathbf{Q}_{12} \\ \mathbf{0} & \mathbf{0} \end{pmatrix}$, where $\mathbf{Q}_{12} = \mathbf{E}_{11}^{-1}\mathbf{E}_{12}$. Then, compute

$$\mathbf{E}_1 = \mathbf{E}_0 - \mathbf{A}_0\mathbf{Q}_0 \quad \text{leading to} \quad \mathbf{E}_1 = \begin{pmatrix} \mathbf{E}_{11} & (\mathrm{I} + \mathbf{A}_{11}\mathbf{E}_{11}^{-1})\mathbf{E}_{12} - \mathbf{A}_{12} \\ \mathbf{0} & \mathbf{A}_{21}\mathbf{Q}_{12} - \mathbf{A}_{22} \end{pmatrix}.$$

Since $\mathbf{E}_1$ is nonsingular, this DAE has a tractability index-1. We, then construct the bases $\boldsymbol{p}_0 \in \operatorname{Im} \mathbf{P}_0$ and $\boldsymbol{q}_0 \in \operatorname{Ker} \mathbf{E}_0$ given by $\boldsymbol{p}_0 = \begin{pmatrix} \mathrm{I} \\ \mathbf{0} \end{pmatrix} \in \mathbb{R}^{n,n_1}$, $\quad \boldsymbol{q}_0 = \begin{pmatrix} -\mathbf{Q}_{12} \\ \mathrm{I} \end{pmatrix} \in \mathbb{R}^{n,n_2}$. And, there respective left inverses is given by $\boldsymbol{p}_0^{*T} = \begin{pmatrix} \mathrm{I} & \mathbf{Q}_{12} \end{pmatrix} \in \mathbb{R}^{n_1,n}$, $\ \boldsymbol{q}_0^{*T} = \begin{pmatrix} \mathbf{0} & \mathrm{I} \end{pmatrix} \in \mathbb{R}^{n_2,n}$. Then we can construct decoupling bases $\hat{\boldsymbol{p}}_0$ and $\hat{\boldsymbol{q}}_0$ such that $\hat{\boldsymbol{p}}_0 \in \operatorname{Ker} \boldsymbol{q}_0^T\mathbf{A}_0^T$ and $\hat{\boldsymbol{q}}_0 \in \operatorname{Ker} \mathbf{E}_0^T$, given by

$$\hat{\boldsymbol{p}}_0^T = \begin{pmatrix} \mathrm{I} & -(\mathbf{A}_{12} - \mathbf{A}_{11}\mathbf{Q}_{12})(\mathbf{A}_{22} - \mathbf{A}_{21}\mathbf{Q}_{12})^{-1} \end{pmatrix}, \quad \hat{\boldsymbol{q}}_0^T = \begin{pmatrix} 0 & \mathrm{I} \end{pmatrix}. \tag{3.8.2}$$

Substituting the above matrix, projector and basis chains into (3.4.6), leads to an equivalent decoupled system of (3.8.1) given by

$$\begin{aligned} \mathbf{E}_{11}\xi_p' &= (\mathbf{A}_{11} - (\mathbf{A}_{12} - \mathbf{A}_{11}\mathbf{Q}_{12})(\mathbf{A}_{22} - \mathbf{A}_{21}\mathbf{Q}_{12})^{-1}\mathbf{A}_{21})\xi_p \\ &\quad + (\mathbf{B}_1 - (\mathbf{A}_{12} - \mathbf{A}_{11}\mathbf{Q}_{12})(\mathbf{A}_{22} - \mathbf{A}_{21}\mathbf{Q}_{12})^{-1}\mathbf{B}_2)u, \\ (\mathbf{A}_{21}\mathbf{Q}_{12} - \mathbf{A}_{22})\xi_q &= \mathbf{A}_{21}\xi_p + \mathbf{B}_2 u, \\ \mathbf{y} &= \mathbf{C}_1^T\xi_p + (\mathbf{C}_2^T - \mathbf{C}_1^T\mathbf{Q}_{12})\xi_q. \end{aligned}$$

Example 3.8.2 Consider a semi-explicit index-2 DAE of the for (2.3.1) with system matrices

$$\mathbf{E} = \begin{pmatrix} \mathbf{E}_{11} & \mathbf{0} \\ \mathbf{0} & \mathbf{0} \end{pmatrix}, \ \mathbf{A} = \begin{pmatrix} \mathbf{A}_{11} & \mathbf{A}_{12} \\ \mathbf{A}_{12}^T & 0 \end{pmatrix}, \ \mathbf{B} = \begin{pmatrix} \mathbf{B}_1 \\ \mathbf{B}_2 \end{pmatrix} \ \text{ and } \ \mathbf{C} = \begin{pmatrix} \mathbf{C}_1 \\ \mathbf{C}_2 \end{pmatrix}. \tag{3.8.3}$$

We assume $\mathbf{E}_{11} \in \mathbb{R}^{n_1 \times n_1}$ and $\mathbf{A}_{12}^T\mathbf{E}_{11}^{-1}\mathbf{A}_{12} \in \mathbb{R}^{n_2 \times n_2}$ are non-singular blocks in order to satisfy the index-2 property and $n = n_1 + n_2$ is the dimension of the DAE. We also assume that the matrix pencil $(\mathbf{E}, \mathbf{A})$ is regular and has atleast one eigenvalue in the finite spectrum. Since the matrix pencil $(\mathbf{E}, \mathbf{A})$ is regular, we can choose admissible projectors

$$\mathbf{Q}_0 = \begin{pmatrix} 0 & 0 \\ 0 & \mathrm{I} \end{pmatrix} \quad \text{and} \quad \mathbf{Q}_1 = \begin{pmatrix} \mathbf{E}_{11}^{-1}\mathbf{A}_{12}(\mathbf{A}_{12}^T\mathbf{E}_{11}^{-1}\mathbf{A}_{12})^{-1}\mathbf{A}_{12}^T & 0 \\ (\mathbf{A}_{12}^T\mathbf{E}_{11}^{-1}\mathbf{A}_{12})^{-1}\mathbf{A}_{12}^T & 0 \end{pmatrix} \tag{3.8.4}$$

to construct a projector and matrix chain leading to a final matrix chain is given by

$$\mathbf{E}_2 = \begin{pmatrix} \mathbf{E}_{11} - \mathbf{A}_{11}\mathbf{E}_{11}^{-1}\mathbf{A}_{12}(\mathbf{A}_{12}^T\mathbf{E}_{11}^{-1}\mathbf{A}_{12})^{-1}\mathbf{A}_{12}^T & -\mathbf{A}_{12} \\ -\mathbf{A}_{12}^T\mathbf{E}_{11}^{-1}\mathbf{A}_{12}(\mathbf{A}_{12}^T\mathbf{E}_{11}^{-1}\mathbf{A}_{12})^{-1}\mathbf{A}_{12}^T & 0 \end{pmatrix}, \tag{3.8.5}$$

since $\mathbf{E}_2$ is non-singular thus, this DAE is indeed an index-2 system. We need to construct an equivalent decoupled system of (3.8.3) in the form (3.5.4) In order to decouple (3.8.3) using the proposed approach, we need to first construct the basis vectors $\boldsymbol{q}_0 \in \operatorname{Ker} \mathbf{E}_0$ and $\boldsymbol{p}_0 \in \operatorname{Im} \mathbf{P}_0$ with their respective left inverses given by

$$\boldsymbol{q}_0 = \begin{pmatrix} 0 & \mathrm{I} \end{pmatrix}^T, \ \boldsymbol{q}_0^* = \begin{pmatrix} 0 & \mathrm{I} \end{pmatrix}^T \quad \text{and} \quad \boldsymbol{p}_0 = \begin{pmatrix} \mathrm{I} & 0 \end{pmatrix}^T, \ \boldsymbol{p}_0^* = \begin{pmatrix} \mathrm{I} & 0 \end{pmatrix}^T. \tag{3.8.6}$$

Using (3.8.4), (3.8.6) and Theorem 2.3.2, we obtain:

$$\mathbf{Z}_{q_0} = \mathbf{E}_{11}^{-1}\mathbf{A}_{12}(\mathbf{A}_{12}^T\mathbf{E}_{11}^{-1}\mathbf{A}_{12})^{-1}\mathbf{A}_{12}^T, \quad \mathbf{Z}_{p_0} = \mathrm{I} - \mathbf{Z}_{q_0}. \tag{3.8.7}$$

We can see that $\mathbf{Z}_{q_0}$ and $\mathbf{Z}_{p_0}$ are projectors as expected. Then, we can obtain their respective vector bases given by $\mathbf{z}_{q_0}$ and $\mathbf{z}_{p_0}$, respectively defined as

$$\hat{\boldsymbol{p}}_0 \in \operatorname{Ker} \boldsymbol{q}_0^T\mathbf{E}_2^T \quad \text{and} \quad \hat{\boldsymbol{q}}_0 \in \operatorname{Ker} \boldsymbol{p}_0^T\mathbf{E}_2^T. \tag{3.8.8}$$

Then finally we can obtain the decoupling projectors using span $\hat{\mathbf{z}}_{p_0} \in \operatorname{Ker} (\hat{\boldsymbol{p}}_0^T\mathbf{E}_2 \boldsymbol{p}_0\mathbf{z}_{q_0})^T$ and span $\hat{\mathbf{z}}_{q_0} \in \operatorname{Ker}(\hat{\boldsymbol{p}}_0^T\mathbf{E}_2\boldsymbol{p}_0\mathbf{z}_{p_0})^T$. Hence substituting (3.8.3)–(3.8.8) into (3.5.4), we obtain the decoupled system of DAE (3.8.3) with a differential and algebraic part.

In the above two examples, we have discussed how to decouple index-1 and 2 systems of DAEs with special structures but in practice we always have DAEs with general structures. The main drawback of the existing approaches such as the use of spectral projectors or Drazin inverses used to the numerical feasibility and extension to varying DAE systems. This was the main motivation of the introduction of the matrix and projector chain approach of decoupling linear constant DAEs as proposed by März [24]. The projector and matrix chain approach was numerically feasible but computationally very expensive since one has to use SVD or alike decomposition in order to construct these projector chain. Recently there has been developments of the fast way of computing these projectors as discussed in the next section.

3.9 Fast Construction of Matrix, Projectors and Basis Chain

In this, section, we discuss how to construct matrix, projector and basis chain in a more efficient way. Fortunately, they have been successful development in efficient construction of such projectors, see [36] for more details. This is briefly discussed below. We discuss a fast way to construct projector $\mathbf{Q}_j$ onto the nullspace of $\mathbf{E}_j$ of large sparse matrix. This approach uses the sparse LU decomposition- based routine from [36], called LUQ. Algorithm 1, illustrates the faster way of constructing matrix and projector chain using Definition 3.1.1.

1: **Inputs:** $\mathbf{E}_0 = \mathbf{E}$ and $\mathbf{A}_0$
2: **Output:** Projector and matrix chain: $\mathbf{E}_j,\ j = 0, \ldots, \mu$ and $\mathbf{Q}_j,\ j = 0, \ldots, \mu - 1$ where $\mathbf{E}_\mu$ is a nonsingular matrix.
3: $j = 0$
4: **while** $\mathbf{E}_j$ is a singular matrix **do**
5: Decompose $\mathbf{E}_j$ such that , $\mathbf{E}_j^T = \mathbf{L}_j \begin{pmatrix} \mathbf{U}_j & 0 \\ 0 & 0 \end{pmatrix} \mathbf{R}_j$, where $\mathbf{L}_j, \mathbf{R}_j \in \mathbb{R}^{n\times n}$ are nonsingular matrices, $\mathbf{U}_j \in \mathbb{R}^{r\times r}$ is a nonsingular upper triangular matrix, r is the rank $\mathbf{E}_j$. Using the LU decomposition- based routine in [36]
6: Compute the LU factorization of $\mathbf{L}_j$ with permutation,

$$\tilde{\mathbf{P}}\mathbf{L}_j = \tilde{\mathbf{L}}_j\tilde{\mathbf{U}}_j,$$

where $\tilde{\mathbf{L}}_j, \tilde{\mathbf{U}}_j, \tilde{\mathbf{P}} \in \mathbb{R}^{n\times n}$, such that $\tilde{\mathbf{P}}$ is the permutation matrix with $\tilde{\mathbf{P}}^T\tilde{\mathbf{P}} = \mathrm{I}$, $\tilde{\mathbf{L}}_j$ and $\tilde{\mathbf{U}}_j$ are lower and upper triangular respectively with unit diagonal entries and a condition number around unity.
7: Set $\mathrm{I}_{n-r} \in \mathbb{R}^{n-r\times n}$ equal to the $n - r$ rows of an $n \times n$ identity matrix.
8: Then, apply the (modified) Gram-Schmidt process, denoted by gs(.), leads to an orthonormal column matrix

$$\Psi_j = \mathrm{gs}((\mathrm{I}_{n-r}\tilde{\mathbf{L}}_j^{-1}\tilde{\mathbf{P}})^T) \in \mathbb{R}^{n\times n-r}, \tag{3.9.1}$$

such that $\mathbf{Q}_j = \Psi_j\Psi_j^T = \mathrm{I}$ forms an orthogonal projector onto $\operatorname{Ker}\mathbf{E}_j$, that is $\operatorname{Im}\mathbf{Q}_0 = \operatorname{Ker}\mathbf{E}_0$ and its complimentary projector $\mathbf{P}_j = \mathrm{I} - \mathbf{Q}_j$.
9: Compute $\mathbf{E}_{j+1} = \mathbf{E}_j - \mathbf{A}_j\mathbf{Q}_j$ and $\mathbf{A}_{j+1} = \mathbf{A}_j\mathbf{P}_j$ from (3.1.1).
10: j=j+1
11: **end while**

Algorithm 1: Fast construction of matrix and projector chain (3.1.1)

We note that the matrix and projector chain obtained using Algorithm 1 can only be used to find the tractability index-μ and for decoupling of index-1 systems. For higher index DAEs one has to extend the algorithm using the theory discussed in Sect. 3.1.1. We observe that the basis chain can also be extracted from Algorithm using steps 3–8 since $\operatorname{Im}\mathbf{P}_j = \operatorname{Ker}\mathbf{Q}_j$. For example $\boldsymbol{q}_0 = \Psi_0 \in \operatorname{Ker}\mathbf{E}_0$.

Chapter 4
Index-aware Model Order Reduction

In this chapter, we discuss the index-aware model order reduction (IMOR) and its invariant the implicit-IMOR(IIMOR) method. We use the decoupled systems (3.2.11) and (3.7.1) to derive the IMOR and IIMOR method respectively.

4.1 Index-aware Model Order Reduction (IMOR)

In this section, we present the Index-aware MOR method which can be abbreviated as the IMOR method. This MOR method was first proposed in [1, 2] for the case of index-1 and -2 DAEs, respectively and its generalization in [7]. The IMOR method reduces both the differential and algebraic parts of decoupled systems (3.2.11) for the equivalent DAE (2.3.1). Conventional MOR methods (ODE MOR Methods) can be used to reduced the differential part, however we shall focus on the moment matching methods. The main motivation of the IMOR method was the need to develop computationally efficient methods which can reduce higher index DAEs. Other MOR methods for DAEs can be found in [16, 35] which are based on the spectral projectors to decouple the DAE before model order reduction. However, the spectral projectors are not sufficiently good tools on appropriate generalizations for variable coefficient linear DAEs and nonlinear DAEs, respectively [25]. This gives our decoupling procedure and the IMOR method an advantage over the existing MOR methods for DAEs since it is based on projector and matrix chain introduced by März [25] which are extendable to variable coefficient linear DAEs and nonlinear DAEs, respectively.

Assuming (2.3.1) is of index-μ with the spectrum of its matrix pair $(\mathbf{E}, \mathbf{A})$ with at least one finite eigenvalue. Then, its equivalent decoupled system is of the form (3.2.11) given by

$$\begin{aligned} \xi_p' &= \mathbf{A}_p \xi_p + \mathbf{B}_p \mathbf{u}, \\ -\mathcal{L}\xi_q' &= \mathbf{A}_q \xi_p - \xi_q + \mathbf{B}_q \mathbf{u}, \\ \mathbf{y} &= \mathbf{C}_p^T \xi_p + \mathbf{C}_q^T \xi_q, \end{aligned} \tag{4.1.1}$$

© Atlantis Press and the author(s) 2016

N. Banagaaya et al., *Index-aware Model Order Reduction Methods*,
Atlantis Studies in Scientific Computing in Electromagnetics 2,
DOI 10.2991/978-94-6239-189-5_4

where $\xi_p \in \mathbb{R}^{n_p}$, $\mathbf{A}_p \in \mathbb{R}^{n_p \times n_p}$, $\mathbf{B}_p \in \mathbb{R}^{n_p}$, $\xi_q \in \mathbb{R}^{n_q}$, $\mathbf{A}_q \in \mathbb{R}^{n_q \times n_p}$, $\mathbf{B}_q \in \mathbb{R}^{n_q \times m}$, $\mathcal{L} \in \mathbb{R}^{n_q \times n_q}$ is a strictly lower triangular nilpotent matrix of index-μ. Next, we discuss why it is important to first decouple a DAE system before applying model order reduction. This is done by first transforming (4.1.1) into frequency domain. Taking the Laplace transform of (4.1.1) and assuming $\xi_p(0) = 0$, since it can be chosen arbitrary, we obtain

$$\mathbf{Y}(s) = \mathbf{H}_p(s)\mathbf{U}(s) + \mathbf{H}_q(s)\mathbf{U}(s) - \mathbf{C}_q^T(\mathrm{I} - s\mathcal{L})^{-1}\mathcal{L}\xi_q(0), \tag{4.1.2}$$

where $\mathbf{H}_p(s) = \mathbf{C}_p^T(s\mathrm{I} - \mathbf{A}_p)^{-1}\mathbf{B}_p$ and $\mathbf{H}_q(s) = \mathbf{C}_q^T(\mathrm{I} - s\mathcal{L})^{-1}\Big[\mathbf{A}_q(s\mathrm{I} - \mathbf{A}_p)^{-1}\mathbf{B}_p + \mathbf{B}_q\Big]$, are the transfer functions of the differential and algebraic parts, respectively and $\mathbf{H}(s) = \mathbf{H}_p(s) + \mathbf{H}_q(s)$. In most literature $\mathbf{H}_p(s)$ and $\mathbf{H}_q(s)$ are called the strictly proper and improper parts of the transfer function $\mathbf{H}(s)$, respectively. In order to derive the reduced-order model using the conventional MOR methods, they always assume vanishing initial conditions, i.e., $\xi(0) = 0$ which leads to the input-output relation $\mathbf{Y}(s) = \mathbf{H}(s)\mathbf{U}(s)$, where $\mathbf{H}(s)$ is the transfer function. Then, $\mathbf{H}(s)$ is approximated such that $\mathbf{H}(s) - \mathbf{H}_r(s)$ is small in the suitable system norm, where $\mathbf{H}_r(s)$ is the transfer function of the reduced-order model. However, from (4.1.2), we can observe that, we can not always have this freedom for the case of DAEs since $\xi_q(0)$ does not always vanish to zero for higher index DAEs. This is only possible for the case on index-1 systems since $\mathcal{L} = 0$, which implies $\mathbf{Y}(s) = \mathbf{H}(s)\mathbf{U}(s)$. Thus assuming vanishing initial condition does not affect index-1 DAE. This the reason why conventional MOR methods lead to accurate reduced-order models for index-1 DAEs and fail for higher index DAEs. Hence the best way to reduce DAEs is by applying model order reduction on the differential and algebraic parts separately. This approach guarantees the accuracy of the reduced-order models. For the case of IMOR method, this done as follows. Separating (4.1.1) into differential and algebraic subsystems given by

$$\xi_p' = \mathbf{A}_p\xi_p + \mathbf{B}_p\mathbf{u}, \tag{4.1.3a}$$

$$\mathbf{y}_p = \mathbf{C}_p^T\xi_p, \tag{4.1.3b}$$

and

$$-\mathcal{L}\xi_q' = \mathbf{A}_q\xi_p - \xi_q + \mathbf{B}_q\mathbf{u}, \tag{4.1.4a}$$

$$\mathbf{y}_q = \mathbf{C}_q^T\xi_q, \tag{4.1.4b}$$

respectively. Then, the differential subsystem (4.1.3) can be reduced using any standard MOR method for ODEs. Considering a Petrov-Galerkin projection by constructing two $\times r$ matrices $\mathbf{V}_p$ and $\mathbf{W}_p$, so that $\mathbf{W}_p^T\mathbf{V}_p = \mathrm{I}$. Then, substituting $\xi_p = \mathbf{V}_p\xi_{p_r}$, into (4.1.3) and left multiplying (4.1.3a) by $\mathbf{W}_p^T$, we obtain

$$\begin{aligned} \xi'_{p_r} &= \mathbf{A}_{p_r}\xi_{p_r} + \mathbf{B}_{p_r}\mathbf{u}, \\ \mathbf{y}_{p_r} &= \mathbf{C}^T_{p_r}\xi_{p_r}, \end{aligned} \tag{4.1.5}$$

where $\mathbf{A}_{p_r} = \mathbf{W}_p^T\mathbf{A}_p\mathbf{V}_p \in \mathbb{R}^{r\times r}$, $\mathbf{B}_{p_r} = \mathbf{W}_p^T\mathbf{B}_p \in \mathbb{R}^{r\times m}$ and $\mathbf{C}_p = \mathbf{V}_p^T\mathbf{C}_p \in \mathbb{R}^{r\times \ell}$, such that $r < n_p$. The transfer function of the reduced-order differential subsystem (4.1.5) is given by $\mathbf{H}_{p_r}(s) = \mathbf{C}^T_{p_r}(s\mathrm{I}-\mathbf{A}_{p_r})^{-1}\mathbf{B}_{p_r}$. We note that the projection matrices $\mathbf{V}_p$ and $\mathbf{W}_p$ determine the subspaces of interest and can be computed in many different ways depending on the model order reduction method used. Next, we seek a reduced-order model of algebraic subsystem (4.1.4) which can be written as

$$-\mathcal{L}_r\xi'_{q_r} = \mathbf{A}_{q_r}\xi_{p_r} - \xi_{q_r} + \mathbf{B}_{q_r}\mathbf{u}, \tag{4.1.6a}$$

$$\mathbf{y}_{q_r} = \mathbf{C}^T_{q_r}\xi_{q_r}, \tag{4.1.6b}$$

where $\mathbf{A}_{q_r} \in \mathbb{R}^{\tau\times r}$, $\mathbf{B}_{q_r} \in \mathbb{R}^{\tau\times m}$ and $\mathbf{C}_{q_r} \in \mathbb{R}^{\tau\times \ell}$, $\tau < n_q$ and its transfer function can be written as $\mathbf{H}_{q_r}(s) = \mathbf{C}^T_{q_r}(\mathrm{I}-s\mathcal{L}_r)^{-1}\Big[\mathbf{A}_{q_r}(s\mathrm{I}-\mathbf{A}_{p_r})^{-1}\mathbf{B}_{p_r} + \mathbf{B}_{q_r}\Big]$. The matrices of the reduced-order model (4.1.6) can be constructed as follows. If we substitute $\xi_p = \mathbf{V}_p\xi_{p_r}$ into (4.1.4a) and rearranging, we obtain

$$\tilde{\xi}_q = \mathcal{L}\tilde{\xi}'_q + \mathbf{A}_q\mathbf{V}_p\xi_{p_r} + \mathbf{B}_q\mathbf{u}, \tag{4.1.7}$$

where $\tilde{\xi}_q$ is the approximation of ξ_q induced by the reduction of ξ_p. Without loss of generality the algebraic variable ξ_q, lies in the subspace $\mathcal{V}_q$ given by

$$\xi_q \in \mathcal{V}_q = \mathcal{K}_\mu(\mathcal{L}, \mathbf{R}_q), \tag{4.1.8}$$

where $\mathbf{R}_q = \Big[\mathbf{B}_q \;\; \mathbf{A}_q\mathbf{V}_p\Big] \in \mathbb{R}^{n_q\times(r+1)m}$, we denote by $\mathbf{V}_q$ an orthonormal basis of $\mathcal{V}_q$ so that $\mathbf{V}_q^T\mathbf{V}_q = \mathrm{I}$. We can then write $\xi_q = \mathbf{V}_q\xi_{q_r}$. Substituting $\xi_q = \mathbf{V}_q\xi_{q_r}$ and $\xi_p = \mathbf{V}_p\xi_{p_r}$ into (4.1.4) leads to a reduced-order algebraic subsystem of the form (4.1.6) with system matrices:

$$\mathcal{L}_r = \mathbf{V}_q^T\mathcal{L}\mathbf{V}_q,\; \mathbf{A}_{q_r} = \mathbf{V}_q^T\mathbf{A}_p\mathbf{V}_p,\; \mathbf{B}_{q_r} = \mathbf{V}_q^T\mathbf{B}_q \;\text{ and }\; \mathbf{C}_{q_r} = \mathbf{V}_q^T\mathbf{C}_q.$$

Thus, the IMOR reduced-order model of (2.3.1) is given by

$$\begin{aligned} \xi'_{p_r} &= \mathbf{A}_{p_r}\xi_{p_r} + \mathbf{B}_{p_r}\mathbf{u}, \\ -\mathcal{L}_r\xi'_{q_r} &= \mathbf{A}_{q_r}\xi_{p_r} - \xi_{q_r} + \mathbf{B}_{q_r}\mathbf{u}, \\ \mathbf{y}_r &= \mathbf{C}^T_{p_r}\xi_{p_r} + \mathbf{C}^T_{q_r}\xi_{q_r}, \end{aligned} \tag{4.1.9}$$

with total dimension $r + \tau \ll n$, where r and τ are the dimension of the reduced-order differential and algebraic parts, respectively. The transfer function of the IMOR reduced model is equal to the sum of the transfer function of the differential and algebraic parts given by $\mathbf{H}_r(s) = \mathbf{H}_{p_r}(s) + \mathbf{H}_{q_r}(s)$. The IMOR method is very accurate and leads to simple reduced order models however its inherited decoupled system is computationally expensive to derive. This limits its application to large scale problems. In the next section, we discuss the computationally cheaper invariant of the IMOR method which is called the implicit-IMOR (IIMOR) method.

4.2 Implicit Index-aware Model Order Reduction (IIMOR)

In this section, we discuss the implicit index-aware model order reduction (IIMOR). The basic idea is still the same as the IMOR method but here we apply model order reduction to (3.7.1) instead of (3.2.11). Assume (2.3.1) is of index-μ, with the spectrum of its matrix pencil has at least one eigenvalue. Then, its equivalent implicit decoupled system can be written in the form (3.7.1) given by

$$\begin{aligned} \mathbf{E}_p\xi_p' &= \mathbf{A}_p\xi_p + \mathbf{B}_p\mathbf{u} \\ -\mathcal{L}\xi_q' &= \mathbf{A}_q\xi_p - \mathcal{L}_q\xi_q + \mathbf{B}_q\mathbf{u}, \\ \mathbf{y} &= \mathbf{C}_p^T\xi_p + \mathbf{C}_q^T\xi_q, \end{aligned} \tag{4.2.1}$$

where $\mathcal{L}$ is a nilpotent matrix of index-μ. $\mathcal{L}_q$ is a non-singular lower triangular matrix with block diagonal matrices for $\mu > 1$. $\xi_p \in \mathbb{R}^{n_p}$, $\xi_q \in \mathbb{R}^{n_q}$, $\mathbf{A}_q \in \mathbb{R}^{n_q \times n_p}$, $\mathbf{B}_q \in \mathbb{R}^{n_q \times m}$ and $\mathbf{C}_q \in \mathbb{R}^{n_q \times \ell}$, $\mathbf{C}_p \in \mathbb{R}^{n_p \times \ell}$. Taking the Laplace transform of (4.2.1) and setting $\xi_p(0) = 0$ since it can be chosen arbitrary, then the output function is given by

$$\mathbf{Y}(s) := \mathbf{H}(s)\mathbf{U}(s) + \mathcal{P}(s), \tag{4.2.2}$$

where $\mathbf{H}(s)$ is decomposed as $\mathbf{H}(s) = \mathbf{H}_p(s) + \mathbf{H}_q(s)$, where $\mathbf{H}_p(s) := \mathbf{C}_p^T(s\mathbf{E}_p - \mathbf{A}_p)^{-1}\mathbf{B}_p$ and $\mathbf{H}_q(s) := \mathbf{C}_q^T(\mathcal{L}_q - s\mathcal{L})^{-1}\left[\mathbf{A}_q(s\mathbf{E}_p - \mathbf{A}_p)^{-1}\mathbf{B}_p + \mathbf{B}_q\right]$ are transfer functions corresponding to the differential part and algebraic parts, respectively and $\mathcal{P}(s) := \mathbf{C}_q^T(\mathcal{L}_q - s\mathcal{L})^{-1}\mathcal{L}\xi_q(0)$. The derivation of the IIMOR method goes as follows. We use the same strategy as in the IMOR method by the first splitting the decoupled system (4.2.1) into separate subsystems as

$$\mathbf{E}_p\xi_p' = \mathbf{A}_p\xi_p + \mathbf{B}_p\mathbf{u}, \tag{4.2.3a}$$
$$\mathbf{y}_p = \mathbf{C}_p^T\xi_p, \tag{4.2.3b}$$

and

$$-\mathcal{L}\xi_q' = \mathbf{A}_q\xi_p - \mathcal{L}_q\xi_q + \mathbf{B}_q\mathbf{u}, \tag{4.2.4a}$$
$$\mathbf{y}_q = \mathbf{C}_q^T\xi_q, \tag{4.2.4b}$$

where (4.2.3) and (4.2.4) are the differential and algebraic subsystems. Then the output equation can be reconstructed using: $\mathbf{y} = \mathbf{y}_p + \mathbf{y}_q$. Next, we derive the reduction procedure for (4.2.3) and (4.2.4), respectively.

4.2.1 Reduction of the Differential Part

The differential subsystem (4.2.3) can be reduced using the standard MOR methods for ODEs such as the moment-matching methods, balanced truncation method, etc. Considering a Petrov-Galerkin projection by constructing two $\times r$ projection matrices $\mathbf{V}_p$ and $\mathbf{W}_p$, so that $\mathbf{W}_p^T\mathbf{V}_p = \mathbf{I}$. Then, substituting $\xi_p = \mathbf{V}_p\xi_{p_r}$, into (4.2.3) and left multiplying (4.2.3a) by $\mathbf{W}_p^T$, we obtain The reduced-order subsystem is obtained by using the approximation $\xi_p = \mathbf{V}_p\hat{\xi}_p$, leading to a reduced-order subsystem:

$$\begin{aligned} \hat{\mathbf{E}}_p\hat{\xi}_p' &= \hat{\mathbf{A}}_p\hat{\xi}_p + \hat{\mathbf{B}}_p\mathbf{u}, \\ \hat{\mathbf{y}}_p &= \hat{\mathbf{C}}_p^T\hat{\xi}_p, \end{aligned} \tag{4.2.5}$$

where $\hat{\mathbf{E}}_p = \mathbf{W}_p^T\mathbf{E}_p\mathbf{V}_p$, $\hat{\mathbf{A}}_p = \mathbf{W}_p^T\mathbf{A}_p\mathbf{V}_p \in \mathbb{R}^{r\times r}$, $\hat{\mathbf{B}}_p = \mathbf{W}_p^T\mathbf{B}_p \in \mathbb{R}^{r\times m}$ and $\hat{\mathbf{C}}_p = \mathbf{V}_p^T\mathbf{C}_p \in \mathbb{R}^{r\times p}$. $\hat{\xi}_p \in \mathbb{R}^r$ is the reduced state vector and $\hat{\mathbf{y}}_p \in \mathbb{R}^\ell$ is the approximated output. Thus the dimension of the differential part is reduced to $r \leq n_p$. The transfer function of the reduced-order model (4.2.5) is given by $\hat{\mathbf{H}}_p(s) = \hat{\mathbf{C}}_p^T(s\hat{\mathbf{E}}_p - \hat{\mathbf{A}}_p)^{-1}\hat{\mathbf{B}}_p$.

4.2.2 Reduction of the Algebraic Part

Substituting $\xi_p = \mathbf{V}_p\hat{\xi}_p$ into (4.2.4), we obtain

$$-\mathcal{L}\xi_q' = \mathbf{A}_q\mathbf{V}_p\hat{\xi}_p - \mathcal{L}_q\xi_q + \mathbf{B}_q\mathbf{u}, \tag{4.2.6a}$$

$$\mathbf{y}_q = \mathbf{C}_q^T\xi_q. \tag{4.2.6b}$$

From (4.2.6a), we can observe that the reduction of the differential part induces a reduction on the algebraic part but the order of the algebraic part is unchanged. In order to reduce the algebraic part, we need to take the following steps. We start from (4.2.6a), which can be written as

$$\mathcal{L}_q\xi_q = \mathbf{N}_q\mathcal{L}_q\xi_q' + \mathbf{b}_q, \tag{4.2.7}$$

where $\mathbf{b}_q = \mathbf{A}_q\mathbf{V}_p\hat{\xi}_p + \mathbf{B}_q\mathbf{u}$ and $\mathbf{N}_q = \mathcal{L}\mathcal{L}_q^{-1}$ is also a nilpotent matrix with the same index-μ as $\mathcal{L}$. Without loss of generality the algebraic variable ξ_q, lies in the subspace $\mathcal{V}_q$ given by

$$\xi_q \in \mathcal{V}_q = \mathcal{K}_\mu(\mathcal{L}_q^{-1}\mathbf{N}_q, \mathcal{L}_q^{-1}\mathbf{R}_q). \tag{4.2.8}$$

We denote by $\mathbf{V}_q$ an orthonormal basis of $\mathcal{V}_q$, so that $\mathbf{V}_q^T\mathbf{V}_q = \text{I}$, and we write $\xi_q = \mathbf{V}_q\hat{\xi}_q$. Substituting $\xi_q = \mathbf{V}_q\hat{\xi}_q$ into (4.2.6) and after simplifying, we obtain a reduced-order algebraic subsystem given by:

$$-\hat{\mathcal{L}}\hat{\xi}_q' = \hat{\mathbf{A}}_q\hat{\xi}_p - \hat{\mathcal{L}}_q\hat{\xi}_q + \hat{\mathbf{B}}_q\mathbf{u}, \tag{4.2.9a}$$

$$\hat{y}_q = \hat{\mathbf{C}}_q^T\hat{\xi}_q, \tag{4.2.9b}$$

with $\hat{\mathcal{L}}_q = \mathbf{V}_q^T\mathcal{L}_q\mathbf{V}_q \in \mathbb{R}^{\tau\times\tau}$, $\hat{\mathcal{L}} = \mathbf{V}_q^T\mathcal{L}\mathbf{V}_q\mathbb{R}^{\tau\times\tau}$, $\hat{\mathbf{A}}_q = \mathbf{V}_q^T\mathbf{A}_q\mathbf{V}_p\mathbb{R}^{\tau\times r}$, $\hat{\mathbf{B}}_q = \mathbf{V}_q^T\mathbf{B}_q\mathbb{R}^{\tau\times m}$ and $\hat{\mathbf{C}}_q = \mathbf{V}_q^T\mathbf{C}_q\mathbb{R}^{\tau\times\ell}$. $\hat{\xi}_q \in \mathbb{R}^\tau$ is the reduced algebraic state vector and $\hat{\mathbf{y}}_q \in \mathbb{R}^\ell$ is the approximated output. Thus the dimension of the algebraic part is reduced to $\tau \leq n_q$. The transfer function of this reduced-order model of the algebraic part (4.2.4) is given by $\hat{\mathbf{H}}_q(s) := \hat{\mathbf{C}}_q^T(\hat{\mathcal{L}}_q - s\hat{\mathcal{L}})^{-1}\left[\hat{\mathbf{A}}_q(s\hat{\mathbf{E}}_p - \hat{\mathbf{A}}_p)^{-1}\hat{\mathbf{B}}_p + \hat{\mathbf{B}}_q\right]$. Thus, combining (4.2.5) and (4.2.9), we obtain the IIMOR reduced-order model of (2.3.1) given by

$$\begin{aligned}
\hat{\mathbf{E}}_p\hat{\xi}_p' &= \hat{\mathbf{A}}_p\hat{\xi}_p + \hat{\mathbf{B}}_p\mathbf{u},\\
-\hat{\mathcal{L}}\hat{\xi}_q' &= \hat{\mathbf{A}}_q\hat{\xi}_q - \hat{\mathcal{L}}_q\hat{\xi}_q + \hat{\mathbf{B}}_q\mathbf{u},\\
\hat{\mathbf{y}} &= \hat{\mathbf{C}}_p^T\hat{\xi}_p + \hat{\mathbf{C}}_q^T\hat{\xi}_q,
\end{aligned} \tag{4.2.10}$$

with total dimension $r + \tau \ll n$, where r and τ are the dimension of the reduced differential and algebraic parts, respectively. The transfer function of the IMOR reduced model is equal to the sum of the transfer function of the differential and algebraic parts given by $\mathbf{H}_r(s) = \mathbf{H}_{p_r}(s) + \mathbf{H}_{q_r}(s)$.

For comparison with other existing MOR method, we can rewrite the IIMOR reduced-order model in descriptor form given by

$$\begin{aligned}
\hat{\mathbf{E}}\xi_r' &= \hat{\mathbf{A}}\hat{\xi} + \hat{\mathbf{B}}\mathbf{u}\\
\hat{\mathbf{y}}_r &= \hat{\mathbf{C}}^T\hat{\xi},
\end{aligned} \tag{4.2.11}$$

where $\hat{\mathbf{E}} = \begin{pmatrix}\hat{\mathbf{E}}_p & 0\\ 0 & -\hat{\mathcal{L}}\end{pmatrix}$, $\hat{\mathbf{A}} = \begin{pmatrix}\hat{\mathbf{A}}_p & 0\\ \hat{\mathbf{A}}_q & -\hat{\mathcal{L}}_q\end{pmatrix}$, $\hat{\mathbf{B}} = \begin{pmatrix}\hat{\mathbf{B}}_p\\ \hat{\mathbf{B}}_q\end{pmatrix}$, $\hat{\mathbf{C}} = \begin{pmatrix}\hat{\mathbf{C}}_p\\ \hat{\mathbf{C}}_q\end{pmatrix}$ and $\hat{\xi} = \begin{pmatrix}\hat{\xi}_p\\ \hat{\xi}_q\end{pmatrix}^T$ for DAEs with a differential part and $\hat{\mathbf{E}} = -\hat{\mathcal{L}}$, $\hat{\mathbf{A}} = -\text{I}$, $\hat{\mathbf{B}} = \hat{\mathbf{B}}_q$, $\hat{\mathbf{C}} = \hat{\mathbf{C}}_q$ and $\hat{\xi} = \hat{\xi}_q$ for DAEs without a differential part and the transfer function of the reduced-order model can be written as $\hat{\mathbf{H}}(s) = \hat{\mathbf{C}}(s\hat{\mathbf{E}} - \hat{\mathbf{A}})^{-1}\hat{\mathbf{B}}$.

4.3 Properties of the IMOR and IIMOR Methods

In this section, we discuss the properties of the IMOR and IIMOR methods. It is clear that the properties of IMOR and IIMOR methods depend on the properties of the standard MOR method for ODEs used to reduce the differential part since the algebraic part is always required to be approximated exactly. For instance, if we use moment-matching methods such as the PRIMA [28] method which preserve passivity under some conditions, also the IMOR and IIMOR can preserve the same properties. This is discussed as follows. Consider a Galerkin projection, that is $\mathbf{V}_p = \mathbf{W}_p$. Then, $\mathbf{V}_p$ can be constructed as follows for the case of IIMOR method. We choose an expansion point $s_0 \in \mathbb{C}\backslash\sigma(\mathbf{E}_p, \mathbf{A}_p)$ and then construct an order-r Krylov-subspace generated by $\mathbf{M}_p$ and $\mathbf{R}_p$ given by:

$$\mathcal{V}_p := \mathcal{K}_r(\mathbf{M}_p, \mathbf{R}_p) = \text{Span}\{\mathbf{R}_p, \mathbf{M}_p\mathbf{R}_p, \ldots, \mathbf{M}_p^{r-1}\mathbf{R}_p\},\ r \leq n_p,$$

where $\mathbf{M}_p := (s_0\mathbf{E}_p - \mathbf{A}_p)^{-1}\mathbf{E}_p$, $\mathbf{R}_p := (s_0\mathbf{E}_p - \mathbf{A}_p)^{-1}\mathbf{B}_p$. Then, $\mathbf{V}_p \in \mathbb{R}^{n_p,r}$ denotes the orthonormal basis matrix of the subspace, $\mathcal{V}_p$, so that $\mathbf{V}_p^T\mathbf{V}_p = \text{I}$. The IIMOR can be constructed as discussed earlier.

Recall the transfer function of the DAE (2.3.1) can be composed as

$$\mathbf{H}(s) = \mathbf{H}_p(s) + \mathbf{H}_q(s), \tag{4.3.1}$$

where $\mathbf{H}_p(s) = \mathbf{C}_p^T(s\text{I}-\mathbf{A}_p)^{-1}\mathbf{B}_p$ and $\mathbf{H}_q(s) = \mathbf{C}_q^T(\mathcal{L}_q - s\mathcal{L})^{-1}\Big[\mathbf{A}_q(s\text{I}-\mathbf{A}_p)^{-1}\mathbf{B}_p + \mathbf{B}_q\Big]$, is the transfer of the differential and algebraic parts, respectively, after doing an implicit decoupling. We use (4.3.1) to show that IIMOR methods also has the same properties as moment matching methods, such as PRIMA method, if PRIMA method is used to reduce the differential part. This is discussed as follows.

(i) **Moment Matching Property**.
After the reduction of the differential part of the decoupled system using the block Arnoldi process, it preserves the first r moments of the differential component $\mathbf{H}_p(s)$ of the decomposed transfer function (4.3.1). This leads to the following Theorem

Theorem 4.3.1 *IMOR methods preserves the moment matching property if and only if the conventional MOR method applied on the differential part preserves the moment matching property.*

Proof If we choose the expansion point as $s_0 = 0$ and assume $\mathbf{A}_p$ is nonsingular. Then the transfer function $\mathbf{H}_p(s)$ of the differential part can be written as $\mathbf{H}_p(s) = \sum_{k=0}^{\infty}\mathbf{h}_p^{(k)}s^k$ where $\mathbf{h}_p^{(k)} = (-1)^k\mathbf{C}_p^T\mathbf{M}_p^k\mathbf{R}_p$ are the (block) moments of $\mathbf{H}_p(s)$, $\mathbf{M}_p = -\mathbf{A}_p^{-1}\mathbf{E}_p$ and $\mathbf{R}_p = -\mathbf{A}_p^{-1}\mathbf{B}_p$. Likewise, the transfer function of the PRIMA reduced-order differential part can be written as $\tilde{\mathbf{H}}_p(s) = \sum_{k=0}^{\infty} \tilde{\mathbf{h}}_p^{(k)} s^k$ where $\tilde{\mathbf{h}}_p^{(k)} = (-1)^k \tilde{\mathbf{C}}_p^T \tilde{\mathbf{M}}_p^k \tilde{\mathbf{R}}_p$ are the (block) moments of $\tilde{\mathbf{H}}_p(s)$, $\tilde{\mathbf{M}}_p = -\tilde{\mathbf{A}}_p^{-1} \tilde{\mathbf{E}}_p$ and $\tilde{\mathbf{R}}_p = -\tilde{\mathbf{A}}_p^{-1} \tilde{\mathbf{B}}_p$. Then $\tilde{\mathbf{C}}_p = \mathbf{V}_p^T \mathbf{C}_p$, $\tilde{\mathbf{A}}_p = \mathbf{V}_p^T \mathbf{A}_p \mathbf{V}_p$, $\tilde{\mathbf{B}}_p = \mathbf{V}_p^T \mathbf{B}_p$. We can observe that $\tilde{\mathbf{h}}_p^{(k)}$ can be written as

$$\tilde{\mathbf{h}}_p^{(k)} = -\mathbf{C}_q^T \mathbf{V}_p \Big[(\mathbf{V}_p^T \mathbf{A}_p \mathbf{V}_p)^{-1} \mathbf{V}_p^T \mathbf{E}_p \mathbf{V}_p \Big]^k (\mathbf{V}_p^T \mathbf{A}_p \mathbf{V}_p)^{-1} \mathbf{V}_p^T \mathbf{B}_p.$$

By construction $\mathbf{V}_p \mathbf{V}_p^T$ is a projector onto $\mathcal{K}_r(\mathbf{M}_p, \mathbf{R}_p)$. Thus it holds $\mathbf{V}_p \mathbf{V}_p^T \mathbf{M}_p^k \mathbf{R}_p = \mathbf{M}_p^k \mathbf{R}_p$, $k = 0, 1, \ldots, r-1$. This in turn implies $\mathbf{V}_p^T \mathbf{M}_p^T \mathbf{R}_p = \tilde{\mathbf{M}}_p^k \tilde{\mathbf{R}}_p$, hence $\tilde{\mathbf{h}}_p^{(k)} = \mathbf{h}_p^{(k)}$, $k = 0, 1, \ldots, r-1$. Next, we can show that the induced reduction on the algebraic part of the DAE also preserves the first r moments of the algebraic component of the transfer function, $\mathbf{H}_q(s)$, which can be written as $\mathbf{H}_q(s) = \sum_{k=0}^{\infty} \mathbf{h}_q^{(k)} s^k$ where $\mathbf{h}_q^{(j)} = \mathbf{C}_q \mathcal{L}_q^{-1} \sum_{j=0}^{\mu-1} \mathbf{N}_q^j \Big[\mathbf{A}_q \mathbf{R}_p + \mathbf{B}_q \Big]$, $j = 0, \ldots, \mu - 1$ are the moments and $\mathbf{N}_q = \mathcal{L}\mathcal{L}_q^{-1}$. By construction $\mathbf{V}_q \mathbf{V}_q^T$ is a projector onto $\mathcal{K}_\mu(\mathcal{L}_q^{-1}\mathbf{N}_q, \mathcal{L}_q^{-1}\mathbf{R}_q)$, where $\mathbf{R}_q = \Big[\mathbf{B}_q \ \ \mathbf{A}_q \mathcal{K}_r(\mathbf{M}_p, \mathbf{R}_p) \Big]$. Thus it holds $\mathbf{V}_q \mathbf{V}_q^T (\mathbf{A}_q \mathbf{M}_p^k \mathbf{R}_p + \mathbf{B}_q) = \mathbf{A}_q \mathbf{M}_p^k \mathbf{R}_p + \mathbf{B}_q$. Then, using the identity $\mathbf{V}_p^T \mathbf{M}_p^T \mathbf{R}_p = \tilde{\mathbf{M}}_p^k \tilde{\mathbf{R}}_p$, it is possible to show that $\tilde{\mathbf{h}}_q^{(k)} = \mathbf{h}_q^{(k)}, \quad k = 0, 1, \ldots, r-1$. □

The above discussion implies that the number of matching moments of the IMOR method depends on the MOR method used to reduce the differential part and the index of the DAE system.

(ii) **Passivity Preservation Property**.
A passive system is one that does not generate energy internally. For an LTI system, (strict) passivity is equivalent to the transfer function being (strictly) positive real. Moreover, if we assume that $\mathbf{E}$ is symmetric and nonnegative definite the necessary and sufficient condition for the system admittance matrix $\mathbf{H}(s)$ to be passive has to satisfy the following theorem.

Theorem 4.3.2 *A rational matrix-valued transfer function $\mathbf{H}(s) \in \mathbb{C}^{m \times m}$ is positive real (strictly positive real) if and only if:*

(1) $\mathbf{H}(s)$ is analytic in $\mathbb{C}^+ = \{s \in \mathbb{C} | \mathrm{Re}(s) > 0\}$;
*(2) $\phi(j\omega) = \mathbf{H}(j\omega) + \mathbf{H}^*(j\omega)$ is positive semi-definite (positive definite) for all $\omega \in \mathbb{R}$ such that $j\omega$ is not a pole of $\mathbf{H}(s)$, where * means the conjugate transpose operation;*
(3) If $j\omega_0$ or ∞ is a pole $\mathbf{H}(s)$, then it is a simple pole and the $m \times m$ residue matrix is positive semi-definite.

Using the above theorem, we can discuss the passivity preservation property of IIMOR method as follows.

$$\mathbf{H}(s) = \mathbf{C}^T(s\mathbf{E} - \mathbf{A})^{-1}\mathbf{B}, \tag{4.3.2}$$

$$= \mathbf{H}_p(s) + \mathbf{H}_q(s), \tag{4.3.3}$$

$$= \mathbf{C}_p^T\mathbf{R}_p(s) + \mathbf{C}_q^T(\mathcal{L}_q - s\mathcal{L})^{-1}\Big[\mathbf{A}_q\mathbf{R}_p(s) + \mathbf{B}_q\Big], \tag{4.3.4}$$

$$= \mathbf{C}_p^T\mathbf{R}_p(s) + \mathbf{C}_q\mathcal{L}_q^{-1}\sum_{j=0}^{\mu-1}\mathbf{N}_q^j\mathbf{N}(s)s^j, \tag{4.3.5}$$

$$= \underbrace{\mathbf{C}_p^T\mathbf{R}_p(s) + \mathbf{M}_0(s)}_{\mathbf{H}_{pr}(s)} + \underbrace{\sum_{j=1}^{\mu-1} s^j\mathbf{M}_j(s)}_{\mathbf{H}_{impr}(s)} \tag{4.3.6}$$

where $\mathbf{R}_p(s) = (s\mathbf{E}_p - \mathbf{A}_p)^{-1}\mathbf{B}_p$, $\mathbf{N}(s) = \mathbf{A}_q\mathbf{R}_p(s) + \mathbf{B}_q$, $\mathbf{M}_0(s) - \mathbf{C}_q^T\mathcal{L}_q^{-1}\mathbf{N}(s)$ and $\mathbf{M}_j(s) = \mathbf{C}_q^T\mathcal{L}_q^{-1}\mathbf{N}_q^j\mathbf{N}(s)$. $\mathbf{H}_{pr}(s)$ is the proper part (bounded as $s \to \infty$) and $\mathbf{H}_{impr}(s)$ the improper part (unbounded as $s \to \infty$) of $\mathbf{H}(s)$. We observe that the transfer function $\mathbf{H}_p(s)$ of the differential part is a strictly proper part of $\mathbf{H}(s)$. Based on Theorem 4.3.2, $\mathbf{H}(s)$ is positive real if and only if $\mathbf{H}_{pr}(s)$ and $\mathbf{M}_j(s)$ are positive real. If we reduced the differential part using PRIMA method [28], it can be proved that the differential part of the reduced order model also preserves positive realness under some conditions. However, in order to ensure that the IMOR methods preserve positive realness for higher index DAEs one need to also prove that $\tilde{\mathbf{M}}_j(s)$ of the reduced-order model is also preserves positive realness which is still an open question.

(iii) **Approximation Error**.
The approximation error of the reduced-order models for DAEs should be computed using the input-output transfer function instead of just the transfer function as it has been the case. Thus, the approximation error of the IIMOR method can be computed as

$$\|\mathbf{Y}(s) - \tilde{\mathbf{Y}}(s)\| \leq \|\mathbf{H}(s) - \tilde{\mathbf{H}}(s)\|\,\|\mathbf{U}(s)\| + \|\mathcal{P}(s) - \tilde{\mathcal{P}}(s)\|, \tag{4.3.7}$$

where $\|\mathbf{H}(s) - \tilde{\mathbf{H}}(s)\| \leq \|\mathbf{H}_p(s) - \tilde{\mathbf{H}}_p(s)\| + \|\mathbf{H}_q(s) - \tilde{\mathbf{H}}_q(s)\|$,
$\|\mathcal{P}(s) - \tilde{\mathcal{P}}(s)\| \leq \|\tilde{\mathbf{C}}_q\tilde{\mathbf{L}}_q\sum_{j=0}^{\mu-1}\tilde{\mathbf{N}}_q^j - \mathbf{C}_q\mathcal{L}_q\sum_{j=0}^{\mu-1}\mathbf{N}_q^j\|\,\|\mathcal{Q}(\mathbf{u}(0)) - \tilde{\mathcal{Q}}(\mathbf{u}(0))\|$ and $\mathcal{Q}(\mathbf{u}(0)) = \mathcal{L}\xi_q(0)$ depends on the input data at $t = 0$. Hence the output-input transfer function of the IIMOR method has a small approximation error if and only if

(a) $\|\mathbf{H} - \tilde{\mathbf{H}}\|$ is small

(b) and $\|\mathcal{P}(s) - \tilde{\mathcal{P}}(s)\|$ is also very small in a suitable norm $\|.\|$. Thus, the IIMOR method is validated more efficiently using the above tools.

(iv) **Stability**.
We have already discussed that the differential part of the decoupled system is an inherited ODE from the DAEs. Hence the stability of the IIMOR method also depends on the MOR method used to reduce the differential part.

We note that, the above properties can also be proved in the same way by simplify replacing $\mathbf{E}_p$ by I_p, that is $\mathrm{I}_p = \mathbf{E}_p$.

Chapter 5
Large Scale Problems

In this chapter, we illustrate the robustness of the IMOR method on large scale problems from real-life applications. These applications include problems from power systems, RLC networks and multibody systems. However these methods can be applied to any application that leads to a linear constant DAE. To reduce the differential part, we use either iterative rational Krylov algorithm (IRKA) or Arnoldi algorithm (PRIMA) to reduce the differential part. Both methods leads to either an index-aware reduced-order model (IROM) or an implicit index-aware reduced-order model (IIROM) if applied to either decoupled system (4.1.1) or (4.2.1), respectively.

Example 5.0.1 This is a SISO power system of dimension $n = 13250$ originating from [30]. It is an index-1 system which can be transformed into its equivalent decoupled system in the form either (4.1.1) or (4.2.1) using the proposed matrix, projector and basis chain approach. Both decoupled systems lead to 1664 differential equations and 11586 algebraic equations. We were able to reduce both the differential and algebraic parts separately using the IMOR methods as illustrated in Table 5.1. We used the PRIMA method to reduce the differential part for the both cases and this lead to a IROM and IIROM of dimension $303 \ll 13250$. In Fig. 5.1, we compare the transfer functions and the phase angle of the reduced-order models (ROMs) with that of the original model. We observe that they coincide and their approximation error is small as shown in Fig. 5.2, however the IMOR method is more accurate. Using $\mathbf{u}(t) = 10\sin(\pi t)$ as the input function, we simulated both ROMs and observed that both lead to accurate output solutions as shown in Fig. 5.3 with small approximation error as shown in Fig. 5.4. Hence the IIMOR methods, were able to reduce this power system accurately and effectively. Thus, these are reliable methods for reducing index-1 DAEs.

© Atlantis Press and the author(s) 2016

N. Banagaaya et al., *Index-aware Model Order Reduction Methods*,
Atlantis Studies in Scientific Computing in Electromagnetics 2,
DOI 10.2991/978-94-6239-189-5_5

Table 5.1 Comparison of the IMOR methods

Method	Reduced ODE	Reduced algebraic	Reduced order
	$r_1 \ll 1664$	$r_2 \ll 11586$	$r_1 + r_1 \ll 13250$
IMOR-PRIMA	151	152	303
IIMOR-PRIMA	151	152	303

Table 5.2 Comparison of the IMOR methods

Method	Reduced ODE	Reduced algebraic	Reduced order
	$r_1 \ll 1998$	$r_2 \ll 1001$	$r_1 + r_1 \ll 2999$
IMOR-PRIMA	301	303	604
IIMOR-IRKA	30	60	90

Table 5.3 Comparison of the computational cost

Grid	Order	Decoupled model			Computational cost (s)	
	n	n_p	k_1	k_2	Implicit method	Explicit method
64 × 64	12159	3969	4095	4095	5521.2	–
60 × 60	10679	3481	3599	3599	3667.6	30653.3
56 × 56	9295	3025	3135	3135	5937.8	8604.0
52 × 52	8007	2601	2703	2703	1574.9	5569.7

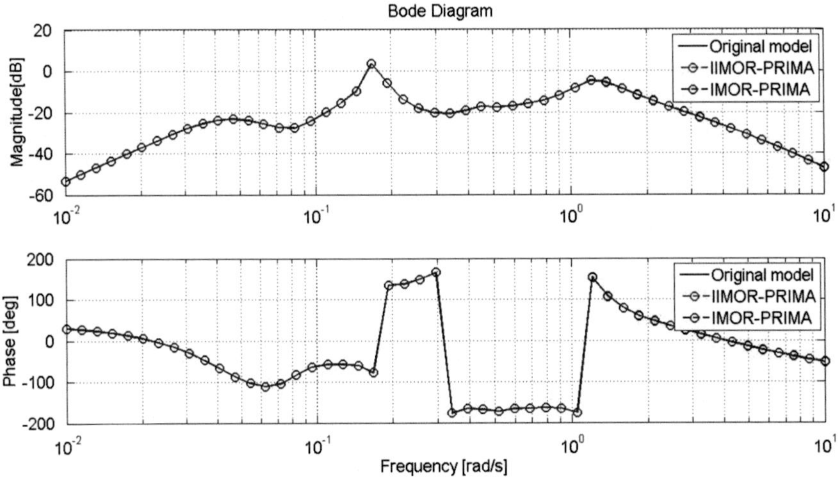

Fig. 5.1 Magnitude and phase plots

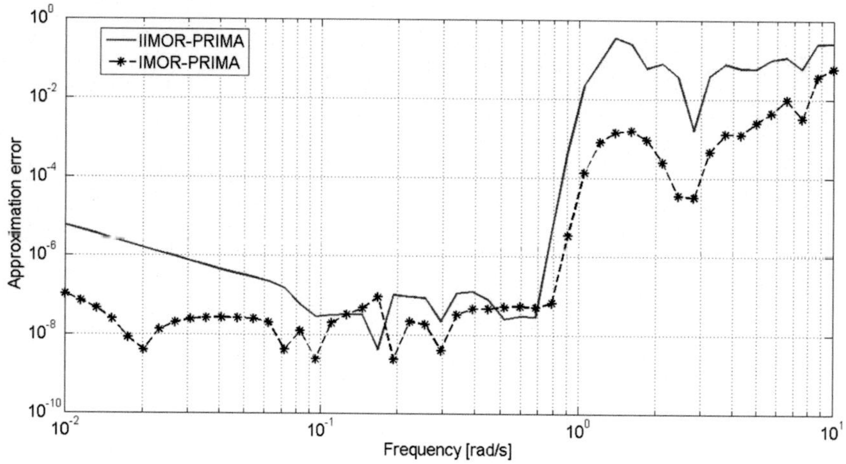

Fig. 5.2 Approximation error in the transfer function

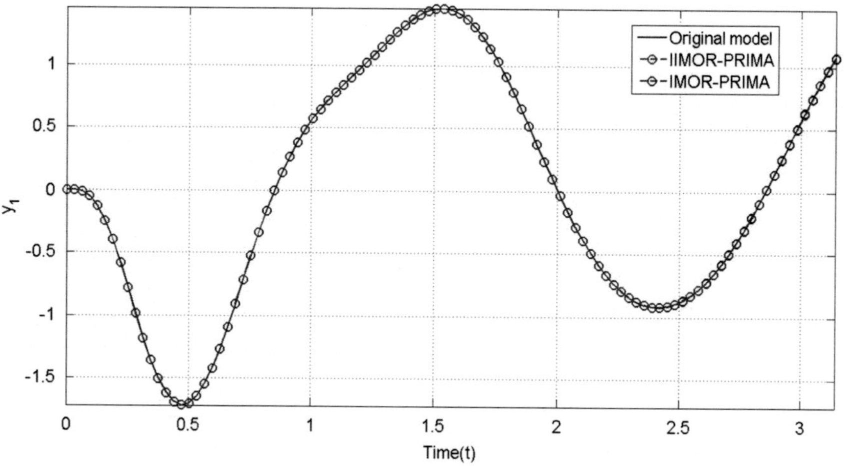

Fig. 5.3 Comparison of the output solution

Example 5.0.2 Consider an RLC circuit in Fig. 5.5 which is modeled using the modified nodal analysis leading to a DAE of the form (2.3.1) written as

$$\begin{pmatrix} \boldsymbol{A}_C \boldsymbol{C} \boldsymbol{A}_C^T & \mathbf{0} & \mathbf{0} \\ \mathbf{0} & \boldsymbol{L} & \mathbf{0} \\ \mathbf{0} & \mathbf{0} & \mathbf{0} \end{pmatrix} \frac{d\boldsymbol{x}}{dt} = \begin{pmatrix} -\boldsymbol{A}_R \boldsymbol{G} \boldsymbol{A}_R^T & -\boldsymbol{A}_L & -\boldsymbol{A}_V \\ \boldsymbol{A}_L^T & \mathbf{0} & \mathbf{0} \\ \boldsymbol{A}_V^T & \mathbf{0} & \mathbf{0} \end{pmatrix} \boldsymbol{x} + \begin{pmatrix} \mathbf{0} \\ \mathbf{0} \\ -1 \end{pmatrix} \boldsymbol{v}, \tag{5.0.1}$$

and $\mathbf{C} = (-1, -1, \ldots, -1)^T)$ with the input function $\mathbf{u}(t) = \boldsymbol{v}(t)$. $n_C = q$, $n_L = q - 1$ and $n_G = q - 1$ are the number of capacitors, inductors and resistors respectively. In this example, we use $\boldsymbol{C}_i = 0.1, i = 1, \ldots, q$, $\boldsymbol{L}_i = 0.5, i = 1, \ldots, q - 1$

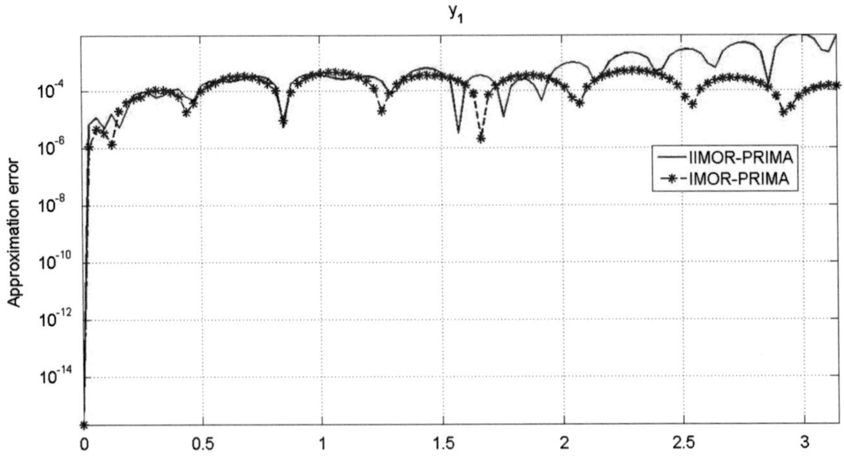

Fig. 5.4 Comparison of error in the output solution

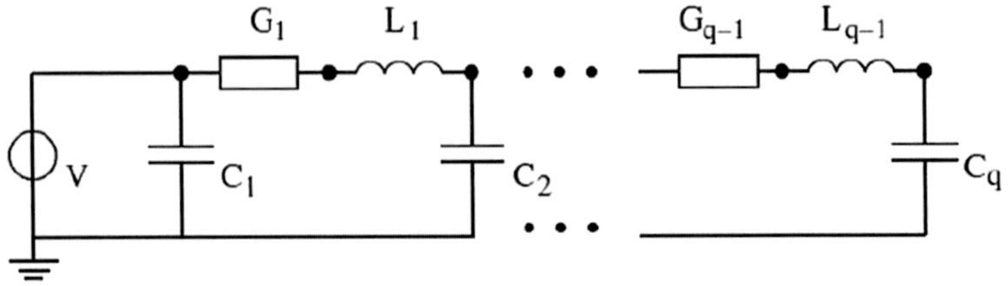

Fig. 5.5 RLC circuit

and $\boldsymbol{G} = 1, i = 1, \ldots, q-1$ as capacitance, inductance and conductances values, respectively and the system dimension $n = 3q - 1$. We observed that this is a SISO index-2 system with a finite spectrum, thus it can be decoupled into either the form (3.6.3) or (3.7.1). We observed that the decoupled system of this DAE system has $2q - 2$ and $q + 1$ differential and algebraic equations, respectively. Using $q = 1000$ leads to a system of dimension 2999 which can be decoupled into $n_p = 1998$ and $n_q = 1001$ differential and algebraic equations, respectively. In Figs. 5.6 and 5.7, shows the sparsity of the respective decoupled system and it is clear that both system are sparse. We intend to reduce the order of this system using IMOR and IIMOR methods by reducing both differential and algebraic parts. We used PRIMA and IRKA to reduce the differential part for the case of IMOR and IIMOR methods, respectively. We observe that the IIMOR method leads to a much smaller ROM than the IMOR as shown in Table 5.2.

In Fig. 5.8, we observe that the transfer function of the ROMs coincides with that of the original model. However, IROM is more accurate at low frequencies while IIROM is more accurate at high frequencies as shown in Fig. 5.9 this is the due to the choice of the interpolation points for the case of IIMOR-IRKA (Fig. 5.9).

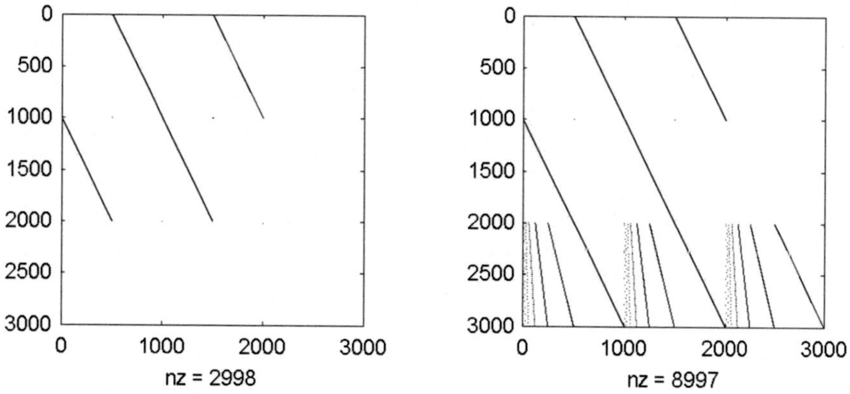

Fig. 5.6 Sparsity of matrix pencil ($\tilde{\mathbf{E}}$, $\tilde{\mathbf{A}}$) of (3.7.3)

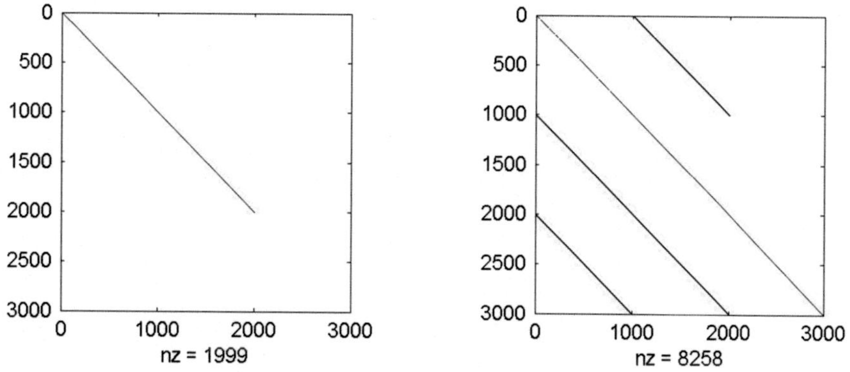

Fig. 5.7 Sparsity of matrix pencil ($\hat{\mathbf{E}}$, $\hat{\mathbf{A}}$) of (3.2.24)

Table 5.4 Comparison of the IMOR methods

Grid	Order	Decoupled model		IIMOR model			IMOR model		
	n	n_p	n_q	r_1	r_2	Time (s)	r	τ	Time (s)
64 × 64	12159	3969	8190	11	12	63.3	–	–	–
60 × 60	10679	3481	7198	11	12	48.8	11	12	13.1
56 × 56	9295	3025	6270	32	33	28.8	32	33	11.7
52 × 52	8007	5406	3599	22	23	19.3	22	23	6.3

Using $\mathbf{u}(t) = \sin(2\pi t)$ as the expansion point, both ROM were simulated and lead to accurate output solutions which coincide with that of the original model as shown in Fig. 5.10. However, the IIMOR-IRKA model is more accurate that IMOR-PRIMA model since it has small approximation error as illustrated in Fig. 5.11.

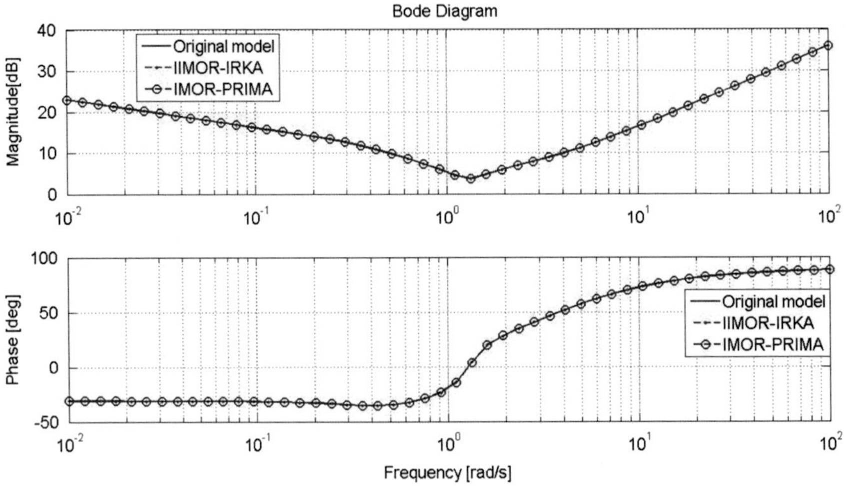

Fig. 5.8 Magnitude and phase plots

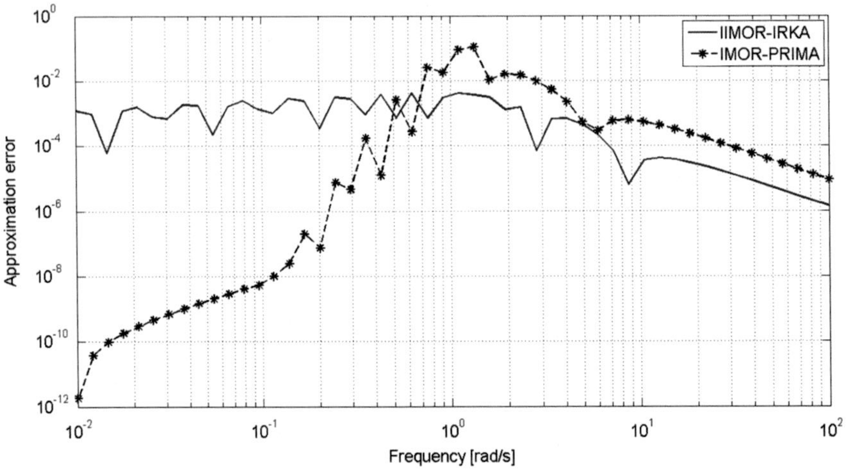

Fig. 5.9 Approximation error in the transfer function

Example 5.0.3 In this example, we apply the IMOR and IIMOR methods on the Semidiscretized Stokes problem as described in Sect. 1.1.2. We performed a spatial discretization of the Stokes equation (1.1.6) on a square domain $\Omega = [0, 1] \times [0, 1]$ by the finite volume method on a uniform staggered grid. This lead to a DAE with system matrices of the form (1.1.7). We observed that this system is a DAE of index-2. In order to compare the computational cost of the implicit and explicit decoupling methods, we carried out experiments on different grid sizes as shown in Table 5.3. From Table 5.3, we see that the finer the mesh the larger the problem becomes. Thus, simulating this system is computationally expensive. We can also

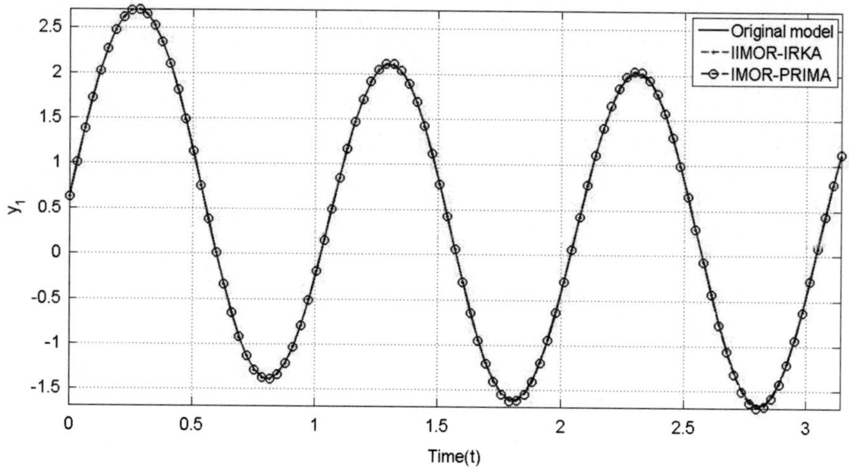

Fig. 5.10 Comparison of the output solution

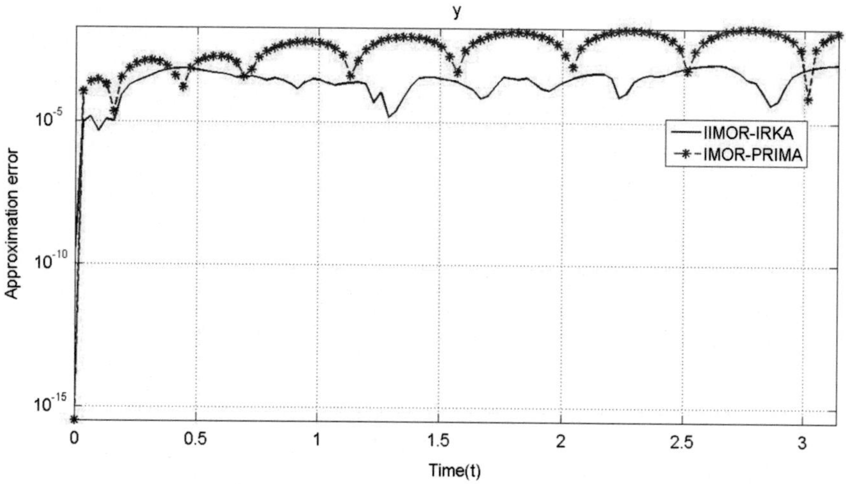

Fig. 5.11 Comparison of error in the output solution

observe that the implicit decoupling method is computationally cheaper than the explicit decoupling method as expected. However, the explicit decoupling method failed to decouple the DAE with grid 64×64 due to very large storage requirements. Next, we reduced the above decoupled Stokes problems using the IMOR and IIMOR methods using different grids. We used PRIMA method to reduce the differential part for all cases and obtained good reduction as shown in Table 5.4. We can observe that the differential and algebraic equations are reduced to order r_1 and r_2, respectively and $r_1 + r_2$ is the order of the reduced-order DAE. We also observe that the IMOR method takes less time than the IIMOR method this is due to the inversion of $\mathcal{L}_q$ but this is small compared to the time it takes to generate the explicit decoupled system as shown in Table 5.3. We used the system matrices from grid 52 × 52 to compare the transfer function and the phase angle of the IMOR model and IIMOR model with that of the original as shown in Fig. 5.12. We can observe that the transfer function and phase angle of the IMOR, IIMOR and original models coincides. However the IMOR model is more accurate than the IIMOR model as shown by the approximation error plot in Fig. 5.13.

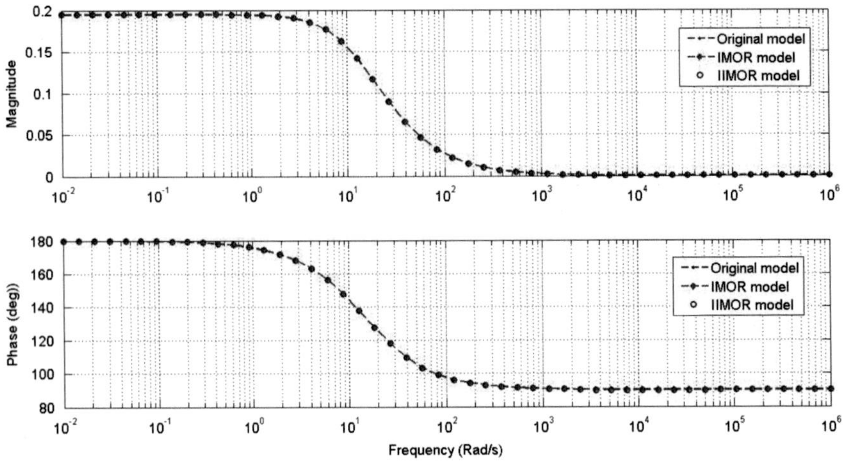

Fig. 5.12 Comparison of the transfer function and phase angle

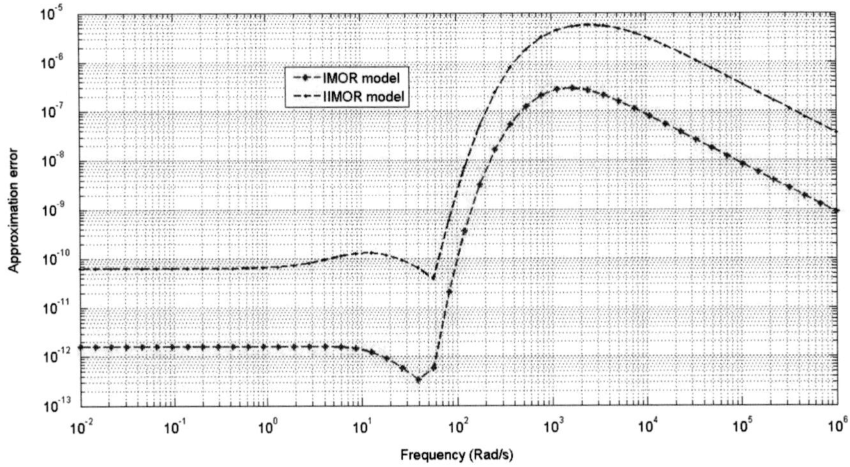

Fig. 5.13 Comparison of the approximation error

We finally compared the solutions of the reduced-order models with that of the original model. From Fig. 5.14, we observe that the solutions of the reduced-order models coincides with that of the original model with a small approximation error as shown in Fig. 5.15. Both reduced model took 10 seconds while the original model took 148 s. Thus the decoupling techniques also make simulation of the full order model computationally very cheap.

Example 5.0.4 In this example, we consider a constrained damped mass-spring system as described in Sect. 1.1.3. This is a DAE of index-3 with its matrix pencil has at least one finite eigenvalue. Substituting $g = 5000$ into (1.1.8) leads to a DAE of

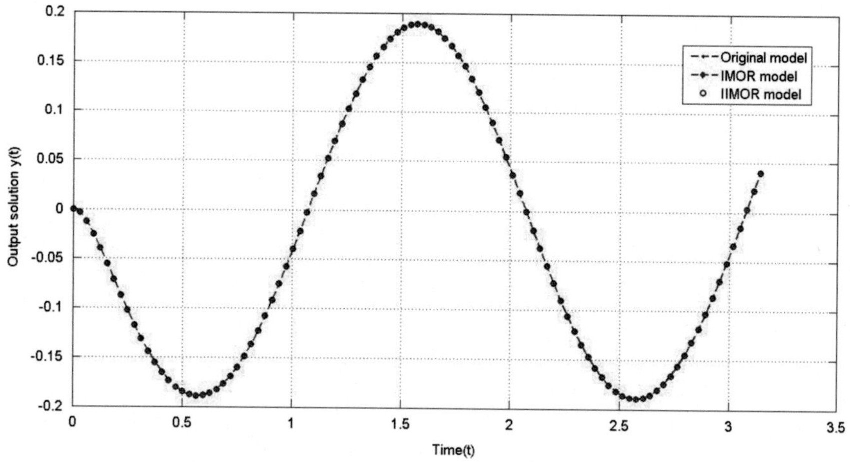

Fig. 5.14 Comparison output solution $y(t)$, $u(t) = \sin(\pi t)$

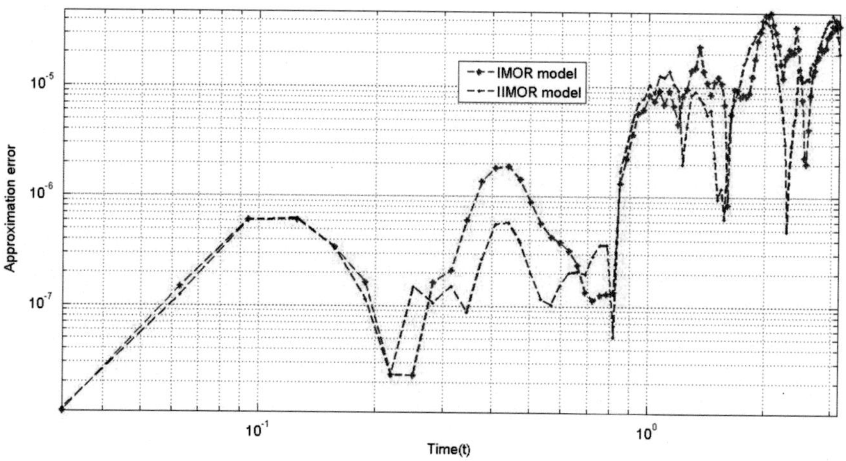

Fig. 5.15 Output solution

Table 5.5 Comparison of the IMOR methods

Method	Reduced ODE	Reduced algebraic	Reduced order
	$r_1 \ll 9998$	$r_2 \ll 3$	$r_1 + r_1 \ll 12001$
IMOR-PRIMA	21	1	22
IIMOR-PRIMA	20	1	21
IIMOR-IRKA	10	1	11

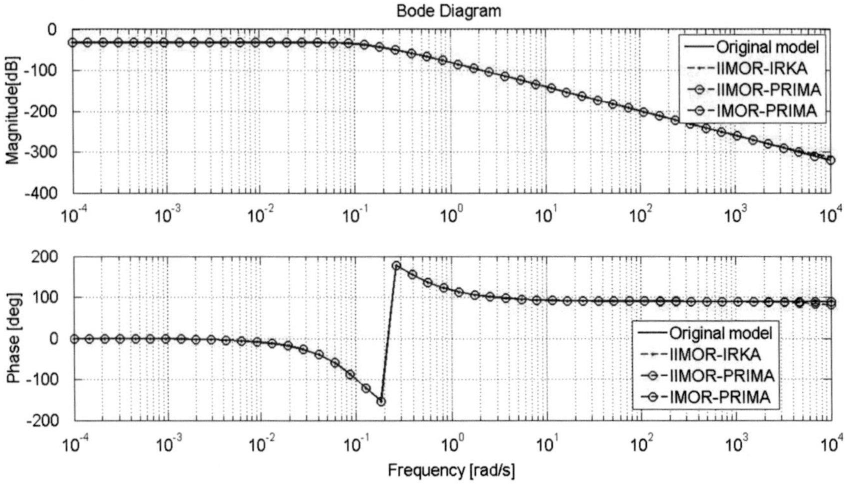

Fig. 5.16 Magnitude and phase plots

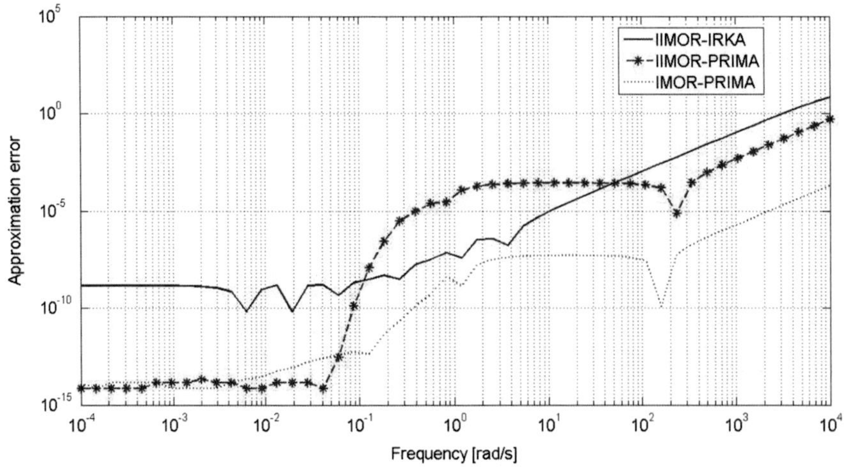

Fig. 5.17 Approximation error in the transfer function

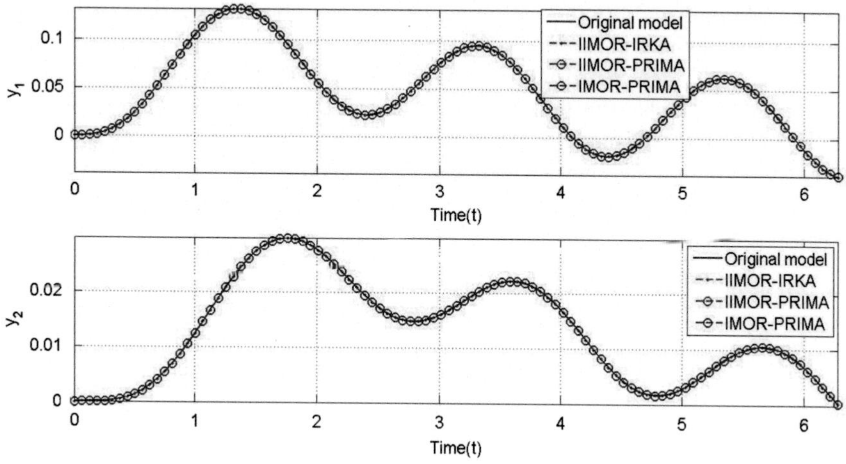

Fig. 5.18 Comparison of the output solution

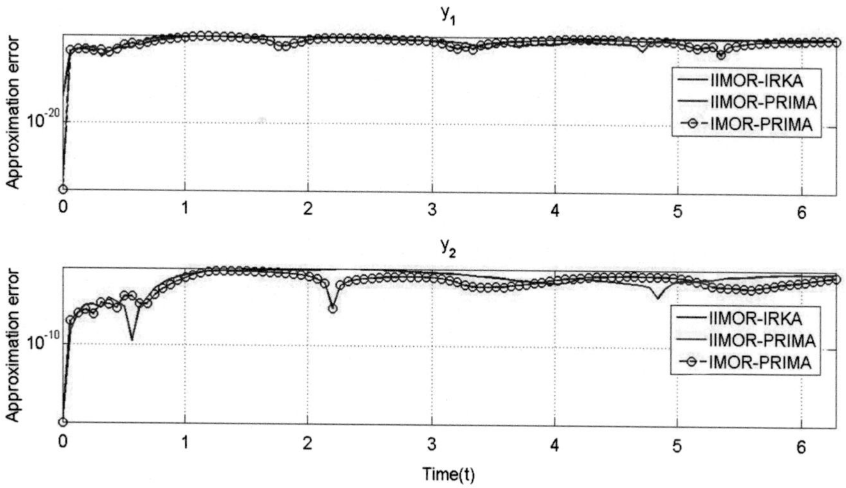

Fig. 5.19 Comparison of error in the output solution

order $n = 10001$ with 1 input and 3 outputs in the form (2.3.1). This DAE system can be rewritten into its equivalent decoupled system of either (4.1.1) or (4.2.1). Both equivalent decoupled system leads to $n_p = 9998$ differential and $n_q = 3$ algebraic equations. We then applied IMOR and IIMOR methods on equivalent decoupled system (4.1.1) and (4.2.1), respectively. For comparison, we used PRIMA and IRKA methods to the reduced the differential parts. In this example, we used both IRKA and PRIMA method to reduce the differential parts for the case of IIMOR method and PRIMA method for the case of IMOR method. Table 5.5, compare their respective reduced dimensions and observe that the IIMOR-IRKA model leads to a much small ROM compared to others. In Fig. 5.16, we can observe that their respective transfer functions and phase angle coincide with that of the original model. However, the models with PRIMA being used to reduce the differential part leads to more accurate transfer function at low frequencies as shown in Fig. 5.17. Using $\mathbf{u}(t) = 10\sin(\pi t)$, we simulated the ROMs and original model, we can observe that there output solutions coincides with that of the original model as shown in Fig. 5.18. The reduced-order models also have the same accuracy since the approximation error of their respective output solutions coincides as shown in Fig. 5.19.

Chapter 6
Conclusion

In this book, two MOR methods for linear constant coefficient DAEs are discussed. These MOR methods are: the Index-aware MOR (IMOR) and Implicit IMOR (IIMOR) methods. They both reduce DAEs of any index by first decoupling it into differential and algebraic parts using projector, matrix and basis chain. However the decoupling procedure of the IIMOR method is computationally cheaper than the IMOR method. Both methods have an attractive property that they preserve the index structure of DAEs. Another interesting feature of these methods is the reduction of the algebraic variables. These methods have been tested on both small and large scale problem from different applications which proves their robustness.

DAEs were not just difficult to reduced but also reduce due to the hidden constraints. Thus, also special techniques were required to simulate them. Fortunately, the implicit and explicit decoupling procedures used to develop the IIMOR and IMOR methods, respectively, can also be used to efficiently simulate DAEs using the standard ODEs integration methods such as Runge-Kutta methods for ODEs. These decoupling procedures are based on the matrix and projector chain approach and are extendable to nonlinear and time varying DAEs. This makes this approach more flexible than the existing decoupling approaches such as the spectral projectors, Drazin inverses, etc. Matrix, projector and basis chains for general structured DAEs are numerically feasible and can be constructed using the computationally cheaper LU based routine discussed in [36]. This same routine can be used to construct bases of these projectors. However, one has to be aware that the numerical computation of the bases for the decoupling may involve serious difficulties because of the accuracy sensitive rank decisions. But it is expected to be profitable if the bases functions can be computed in a robust way, for example some applications such as the electrical network problems which are modeled using the incidence matrices, we recommend to use the incidence matrices to construct these bases instead of using singular matrices which may be ill-conditioned. For some applications with special structures they can also be constructed explicitly.

© Atlantis Press and the author(s) 2016

N. Banagaaya et al., *Index-aware Model Order Reduction Methods*,
Atlantis Studies in Scientific Computing in Electromagnetics 2,
DOI 10.2991/978-94-6239-189-5_6

References

1. G. Alí, N. Banagaaya, W.H.A. Schilders, C. Tischendorf, Index-aware model order reduction for differential-algebraic equations. Math. Comput. Model. Dyn. Syst.: Methods Tools Appl. Eng. Relat. Sci. **20**(4), 345–373 (2013)
2. G. Ali, N. Banagaaya, C. Tischendorf, Index-aware model order reduction for index-2 differential-algebraic equations with constant coffiecients. SIAM J. Sci. Comput. **35**(3), A1487–A1510 (2013)
3. A.C. Antoulas, *Approximation of Large-Scale Dynamical Systems* (SIAM, Philadelphia, 2005)
4. N. Banagaaya, Index-aware model order reduction methods. PhD thesis, Eindhoven University of Technology, Eindhoven, Netherlands, 2014
5. N. Banagaaya, G. Alí, W.H.A. Schilders, Implicit-IMOR method for index-1 and index-2 linear constant DAEs. External Report, CASA Report, No. 14–10). Eindhoven: Technische Universiteit Eindhoven, 28 pp (2014). http://www.win.tue.nl/analysis/reports/rana14-10.pdf
6. N. Banagaaya, W.H.A. Schilders, Simulation of electromagnetic descriptor models using projector. J. Math. Ind. **3**(1), 1–18 (2013)
7. N. Banagaaya, W.H.A. Schilders, Index-aware model order reduction for higher index DAEs, in *In Progress in Differential-Algebraic Equations*, ed. by S. Schöps, A. Bartel, M. Günther, E.J.W. ter Maten, P.C. Müller (Springer, Berlin, 2014), pp. 155–182
8. S. Baumanns, Coupled electromagnetic field/circuit simulation: modeling and numerical analysis. PhD thesis, Universität zu Köln, Germany, 2012
9. P. Benner, M. Hinze, E.J.W. ter Maten (eds.), *Model Order Reduction for Circuit Simulation* (Springer, Dordrecht, 2011)
10. S.L. Campbell, *Singular Systems of Differential Equations. I* (Pitman, San Fransisco, 1980)
11. L. Dai, *Singular Control Systems. Lecture Notes in Control and Infomation Sciences* (Springer, New York, 1989)
12. G.-R. Duan, *Analysis and Design of Descripror Linear Systems*, vol. 23 (Springer, New York, 2010)
13. F.D. Freitas, N. Martins, S.L. Varrichio, J. Rommes, F.C. Véliz, Reduced-order tranfer matrices from network descriptor models of electric power grids. IEEE Trans. Power Syst. **26**, 1905–1919 (2011)
14. F.D. Freitas, J. Rommes, N. Martins, Gramian-based reduction method applied to large sparse power system descriptor models. IEEE Trans. Power Syst. **23**(3), 1258–1270 (2008)
15. S. Grundel, L. Jansen, N. Hornung, P. Benner, T. Clees, C. Tischendorf, Model order reduction of differential algebraic equations arising from the simulation of gas transport networks. Technical Report MPIMD/13-09 (2013)

© Atlantis Press and the author(s) 2016

N. Banagaaya et al., *Index-aware Model Order Reduction Methods*,
Atlantis Studies in Scientific Computing in Electromagnetics 2,
DOI 10.2991/978-94-6239-189-5

16. S. Gugercin, T. Stykel, S. Wyatt, Model reduction of descriptor systems by interpolatory projection methods. SIAM J. Sci. Comput. **35**(5), B1010–B1033 (2013)
17. B. Hassen, Linear differential algebraic equations, Seminarbericht. Technical Report 92–1, Humboldt-Univ. Berlin, Fachbereich Mathematik (1992)
18. C.W. Ho, A.E. Ruehli, P.A. Brennan, The modified nodal approach to network analysis. IEEE Trans. Circuits Syst. CAS **22**(6), 504–509 (1975)
19. G. Kron, *Tensor Analysis of Networks* (General Electric Series, New York, 1939)
20. R. Lamour, R. März, C. Tischendorf, Projector based treatment of linear constant coefficient DAEs. Technical Report 11–15, Humboldt Universität zu Berlin, Germany, Department of Mathematics (2011)
21. R. Lamour, V. Mehrmanann, *Differential-Algebaric Equations (Analysis and Numerical Solution)* (EMS, Zürich, 2006)
22. G. Lassaux, High-fidelity recuded-order aerodynamics models: application to active control of engine inlets. Master's thesis, Departmentt of Aeronutics and Astronautics, MIT, USA, 2002
23. G. Lassaux, K. Wlillcox, Model reduction of an actively controlled supersonic diffuser, in ed. by P. Benner, V. Mehrmann and D.C. Sorensen, *Dimension Reduction of Large Scale Systems, Lecture notes in Computational Science and Engineering* (Springer, Berlin, 2005), pp. 357–361
24. R. März, Numerical methods for differential algebraic equations. Cambridge University Press, Acta Numerica **21**(5), 141–198 (1992)
25. R. März, Canonical projectors for linear differential algebaric equations. Comput. Math. Appl. **31**(4–5), 121–135 (1996)
26. W.J. McCalla, *Fundamentals of Computer Aided Circuit Simulation* (Academic Publication Group, Kluwer, Dordrecht, 1988)
27. V. Mehrmann, R. Nabben, E. Virnik, Generalisation of the perron-frobenius theory to matrix pencils. ScienceDirect, Linear Algebra Appl. **428**(1), 20–38 (2008)
28. A. Odabasioglu, M. Celik, L.T. Pileggi, PRIMA: passive reduced-order interconnect macromodeling algorithm. IEEE Trans. Comput.-Aided Des. Integr. Circuits Syst. 17(8), 645–654 (1998)
29. R. Riaza, *Differential-Algebraic Systems: Analytical Aspects and Circuit Applications* (World Scientific Publishing Co., Pte. Ltd, New Jersey, 2008)
30. J. Rommes, N. Martins, Computing large-scale system eigenvalues most sensitive to parameter changes, with applications to power system small-signal stability. IEEE Trans. Power Syst. **23**(2), 434–442 (2008)
31. W. Schilders, H. Van der Vorst, J. Rommes (eds.), *Model Order Reduction: Theory, Research Aspects and Applications* (Springer, Berlin, 2008)
32. S. Schulz, Four lectures on differential algebraic equations. Technical report, Humboldt Universität zu Berlin, Germany (2003)
33. T. Stykel, Analysis and numerical solution of generalized Lyapunov equations. PhD thesis, Technischen Universität Berlin, Germany, 2002
34. T. Stykel, Anaylsis and Numerical solution of generalized Lyapunov Equations. PhD thesis, Technical University Berlin, 2002
35. V. Mehrmann, T. Stykel, Balanced truncation model reduction for large scale systems in descriptor form, in *Dimension Reduction of Large Scale Systems, Lecture notes in Computational Science and Engineering*, vol. 45. (Springer, Berlin, 2005), pp. 83–115
36. Z. Zhang, N. Wong, An efficient projector-based passivity test fot descriptor systems. IEEE Trans. Comput. Aided Des. Integr. Circuits Syst. 29, 1203–1214 (2010)
37. Z. Zhang, N. Wong, Canonical projector techniques for analyzing descriptor systems. Int. J. Control Autom. Syst. **12**(1), 71–83 (2014)